# HOW TO TUMBLE ROCKS INTO GEMS

## SECRETS OF THE PROS REVEALED

In this revised second printing of *How to Tumble Rocks into Gems, Secrets of the Pros Revealed,* we have added section numbers to the pages. This is make it easier for the reader to cross reference the *contents section (iii--vii)* with book *pages* .

Sections are in *NUMERICAL ORDER AND LOCATED AT THE TOP INSIDE OF EACH PAGE.*

Examples: *Loading the Vibratory Tumbler.......* Section #9

Turn pages and refer to the section numbers at the top inside location on the page.

You may still find information in alphabetical order by using the index section in the rear of the book...page #99.

**DISCLAIMER**

Liability is herewith disclaimed for any damages, real or consequential, resulting from the use of information contained in this publication. Although information has been obtained from sources which we believe to be reliable, the accuracy and validity contained herein is neither guaranteed or warranted in any manner whatsoever. It is the obligation of the reader and/or user to determine the suitability and fitness of purpose for any application in which this information is used or applied.

# HOW TO TUMBLE POLISH ROCKS INTO GEMS

## SECRETS OF THE PROS™ REVEALED


ISBN 0-9765603-1-3



**First Printing, January 1995**
**Second (Revised) Printing, January 1996**
**Third Printing, April 1997**
**Fourth Printing, October 1998**
**Fifth Printing, March 2000**
**Sixth Printing, June 2001**
**Seventh Printing, September 2002**
**Eighth Printing, March 2003**
**Ninth Printing, July 2004**
**Tenth (Revised) Printing Third Edition, August 2005**
**Eleventh (Revised) Printing Third Edition, August 2006**
**Twelfth Printing Third Edition, July 2007**

Published by Secrets of the Pros
Printed and bound in the United States of America
Secrets of The Pros % E. & J. Smith

## THE OBJECTIVE OF THIS BOOK

It will be the objective to try to explain, instruct, and to guide the reader through the basic processes that in most instances--whether the field is lapidary, industrial, or commercial--parallel one another in basic lapidary Grinding, Sanding and Polishing cycles. During the author's 40 plus years "love affair" with tumbling literally thousands of pounds of materials have been processed in experimentation. Accurate processing records were maintained during this period and in this book, with secrets of the pros revealed, the authors life-long knowledge in the field of tumbling is set forth for the use and enlightenment of those persons who are inclined to pursue the hobby of tumbling, and for those who delve further into the commercial aspects of the trade.

Literally hundreds of processes have been implemented, both experimental, and trial and error processing. Hundreds and hundreds of samples of almost every conceivable material from around the globe were received by the author for processing or formula development for customers and users of the equipment. Technical terms are held to a minimum, and not only does the author impart the how, but, in many instances, explains the why, both in Rotary Barrel tumbling and in Vibratory Finishing.

This book sets forth the instructions, guide lines and methods learned from experience. From this we can truthfully expound; "IT IS DIFFICULT, IF NOT ALMOST IMPOSSIBLE TO IMPART FACTUAL KNOWLEDGE, AND INSTRUCTION, WITHOUT EVER HAVING EXPERIENCED THE PROBLEM." We learn profoundly from our mistakes, which makes the successes all the more worthwhile!
Good Luck and Good Tumbling! Edward E. Smith, Author.

**SECTION** # *Contents* **PAGE**

**SECTION** **Contents** **PAGE**

# Contents

SECTION

# Contents

PAGE

SECTION **Contents** PAGE

Photo by Steve Smith

Author Edward E. Smith explains the workings of ultra-sonics and surface measuring equipment to students during 1978 Geo-Sonics Seminar.

# PREFACE

This book is dedicated to my wife Lois, my father the late E.N Smith, founder of Geode Industries Inc., and to my family, without whose continued support this book would not have been possible. To those in the trade who encouraged me to share the knowlege gained during 40 odd years of experience with tumbling. Also to former employees of Geode Industries Inc. and Geo-Sonics, (especially Bill McCabe, Audrey Williamson, and Jeri Lyles) who so faithfully helped in experiments and record keeping.

## HOW IT ALL BEGAN [1933]

As far back as I can recall, there has always been present a basic inherent interest in mother earth...possibly this is basic in all of us in one degree or another. Sort of an uncontrollable urge to seek out things that are new and different...to observe and hopefully appreciate the natural beauty of this world. Discovering that much of the creators handiwork can be revealed through just a small amount of effort on our part.

Many of the treasures in pattern, symmetry, color and design are disguised under the plainest of wrappers. If we are to enjoy this beauty, then we must persue it Take fishing for instance...the real purpose is not to return with the catch each and every time...rather what we really remember, long after the savory taste of the catch has departed, is the trip, the scenery, the views, companionship....just getting away and enjoying nature.

Its like that with rock hunting and collecting. The objective may be to collect and produce something beautiful from our finds. That's lapidary! Sharing new discoveries with others, enjoying nature, the lore of the unknown, are extras.

So it was with myself…at age four…I can barely recall the early trips to the north shore of Lake Superior, however there are some experiences which seem to loom more prominently than others. Tent camping, picking blueberries, (including the savory taste of fresh baked pie)…it's still my favorite.

Too young to tramp the fishing trails with the men, I applied my time to roaming the sandy and pebble covered beaches accompanied by aunts and other campers. This wasn't all bad however as I found to my liking there were curious looking pebbles in the crystal clear waters of Lake Superior. Even the 52 degree temperature of the lake didn't deter me from the pursuit of collecting rocks. Some seemed different in several respects...more pattern, irregular in shape, brighter in color, bands, lines, some with "eyes" and even mossy patterns. This now was a real challenge, and I decided to give full pursuit.

And so it was with the beaches of Grand Marias, each year returning (older of course) and collecting for the next 5 years the odd looking stones with myriads of delightfully pleasing displays of color and design. My jelly jars were filled with Lake Superior Agates and Thompsonites. Each time I rolled the jars the stones tumbled to show even more magnified beauty. There must be a way, I thought, to reveal all of the color and pattern concealed below the knobby and pitted surface.

There was…it would be revealed to me 10 years later. Rock Tumbling [using tumbling machines to remove the surface, exposing, and preserving the beauty] would become a life-long pursuit of turning Rocks into Gems.

# INTRODUCTION TO TUMBLING

If you have pondered about those stones and rocks you have been collecting on vacations, along the road, while at the beach, or when enjoying the out-of-doors..stored in jars, cans, boxes or piling up in the corner of the garage...don't put it off another day..go get them, purchase a tumbler and lets start tumbling!

---

*"how to tumble everything from Apache Tears to Zebra Agate...Jewelry Castings to...Shell Casings..Fossils to Sea Shells...Carvings to Cabochons"*

---

In this book we're going to cover everything from; Apache Tears to Zebra Agate...Jewelry Castings to...Brass Shell Casings... Fossils to Sea Shells and from Carvings to Cabochons. *I never said it was going to be easy*...just that we'll make it PLAIN AND SIMPLE. We'll discuss Gemstones, Jewelry Castings, Fossils, Pre-Forms & Cabochons.. First off, lets learn about tumbling machines, the objectives we want to accomplish through the tumbling process and the procedures, abrasives, and compounds involved.

My first tumbler back in '52 was real generic, there weren't many commercial machines on the market, and what were available were just plain barrel type tumblers. Here's the recipe I used:

<u>THE RECIPE</u>

1 Airplane tire [well used but not blown or patched]

1 Rim [cut out on one edge for easy loading and removal

1 Inner tube to match tire

1 2x12 6 ft. long [used but not split or cracked]

1 2x12 4 ft. long [base plate]

1 Steel shaft 1" x 10" [Welded to plate on Steel Rim]
2 Bearings [Mounted on opposite end of rim and shaft]
1 Pulley [Mounted on open of rim and shaft]
1 Washing Machine motor & belt [Used but oprable w/ Pulley

As you can surmise from the description, capacity and reduction of the long times normally required (4-5 weeks) was the objective.

THE LOAD: 40 LBS LAKE SUPERIOR AGATES; 4 LBS #80 GRIT; WATER TO COVER TO LOAD.

Week #1 ended in Disaster, little stock removal or smoothing ,but "whoa"...lots of fracturing and chipping of the edges..what had gone wrong...the scant directions were followed carefully. Load the tire with 40-50 lbs of rock, partially inflate the tube to contain splash, add 1 pound of grit for every 10 lbs of stone, water to cover and "turn 'er on and let 'er run" for a week. Here's where we need to aquire some basics about tumbling machines and how important it is to load the tumbler correctly.

BASIC TYPE OF TUMBLING MACHINES

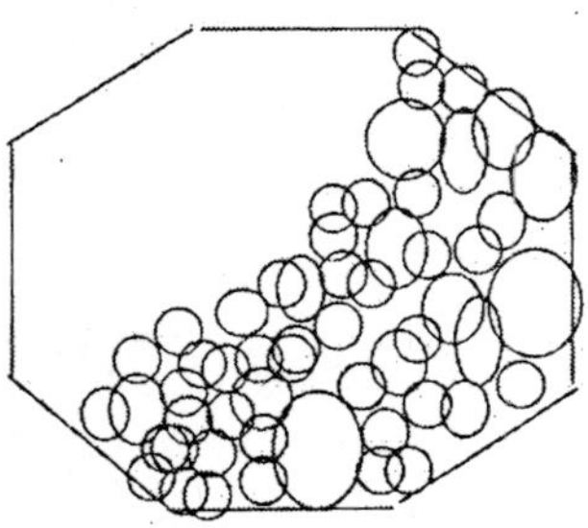

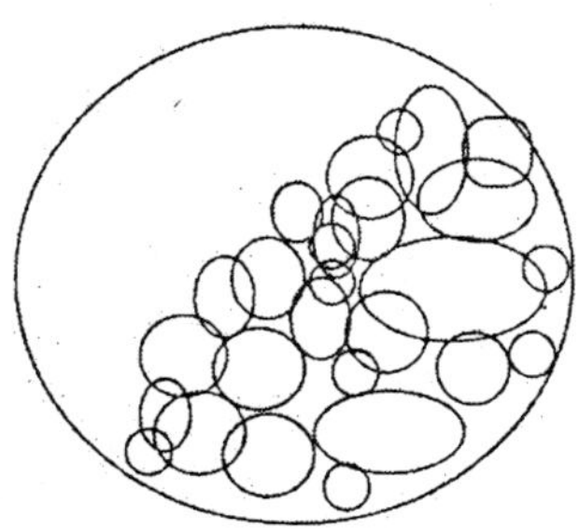

Rotary units are usually round or hexagon shaped barrels, or round barrels with hex flats molded into the liner. Speed of Rotation is very important for any given diameter and usually ranges from 10 to 35 rpm. Too slow--and the load will slide along the bottom with little or no tumbling action. Too fast--and the stones will go up and over the top [cascading] with chipping, spalling and fracturing, especially of the edges.

# TYPES OF TUMBLING MACHINES

## ROTARY

There is no need for any real concern over the speed of the barrel...most all machines that are on the market today have set speeds that are correct for the diameter of the barrel. Select your equipment from a reputable dealer and follow the manufacturer's instructions. Single barrel models are the most economical, with capacities starting at about 1½ pounds to around 50 lbs.

---

*"...all types...have advantages and disadvantages ...make your decision based on costs...quantity ...materials...end results."*

---

Single barrel small capacity units [1½ to 4 lbs] and multiple barrel types are the least expensive, well suited for beginners and youngsters alike. Especially where smaller quantities of materials are to be processed, and the physical size and shape of the material is small, averaging form 3/8" to 1½".
Larger barrel tumblers [6-12 lbs.] considered to be more in the professional class, will process larger loads, handle larger sized materials [1/2" to 3"], and many times longer pieces. Barrel tumblers with 12 to 50 lb. capacity, considered commercial, have advantages in size, quantity and slightly shorter processing cycles.

In general, barrel tumblers load quickly, easily, require a minimum of attention, and produce satisfactory results on most Gem Materials, and metals. Rotary barrel tumblers also tend produce more rounded surfaces, blend contours, and are easiest to operate on materials from 6 to 7½ in hardness on the MOH'S scale.

# TYPES OF TUMBLING MACHINES

## VIBRATORY

There are several types of Vibratory Machines that basically produce both vibratory and rotary action, simultaneously. Vibration frequency range from 1500 to 3600 vibrations per minute, plus rotary action in vertical and horizontal planes. Criteria for selection of vibratory finishers is basically similar to making a decision for rotary machines. Capacities from 4 lbs to several hundred pounds are available, based on requirements. Quantity, type, shape, hardness of materials, age, physical condition of the operators, time element, and desired end results should be factors involved in the decision.

Initial costs for vibratory finishers are somewhat higher, however reduced tumbling times, less abrasives, and in electronic models, considerably reduced power consumption, can all help to offset price disadvantages. Multiple container units offer greater capacities, operational economy, and multiple operations at the same time are additional advantages. Interchangeable hoppers on some models offer additional economy.

---

*"some vibratory finishers produce results in 5-10 days as opposed to 4--6 weeks with most barrel types"*

---

Vibratory finishers are either mechanical or electronic drives. Mechanical units uise motors, shafts, belts, pulleys, and eccentrics. Electronic units employ fewer parts, developing energy magnetically, without motors. Open type containers load quickly and offer ease of inspection. Operation is different with contents processed with a pancake batter like slurry. Results are usually produced in 5-10 days as opposed to 4-6 weeks for most rotary units.

# WHAT TO LOOK FOR AND WHERE

## LOCATIONS

Tumbling materials can be found most anywhere the most lucrative being, creeks, streams, lakes, beaches, road cuts, gravel pits, quarries, deserts, mountains, and ocean beaches. Even your own back yard. Who knows what gem materials lurk in that old gravel pile, flower bed, or decorative motif. Much of the gravel used in landscaping comes from sand pits and ancient stream beds, sometimes pumped to surface from glacial formations. In any event the natural processes of nature have pre-tumbled much of this material in ancient lakes, oceans, streams and glaciers, formed thousands, even millions of years ago. In some instances, transporting stones, hundreds of miles from their original origin. Though this natural process the stones have been rounded and smoothed, making selection by color, pattern and quality, a much easier process. Almost every conceivable type of gem material can be recovered from this material...ideal for processing in tumblers, either rotary barrel or vibratory.

## MATERIALS

Materials that appear different..Shiny as opposed to Dull, Translucent as opposed to Opaque, or Opaque with color and or pattern. If stones are wet when picked up, turn dull and dry almost immediately, chances of polishing are slim. These could be decorative, but not tumbling materials. If collecting a dry area, take along a spray bottle filled with water, pre-select tumbling materials, from decorative or perform this operation later. Large sized materials will probably have to be reduced by crushing or breaking.

# PLAIN AND SIMPLE

## BEGINNING

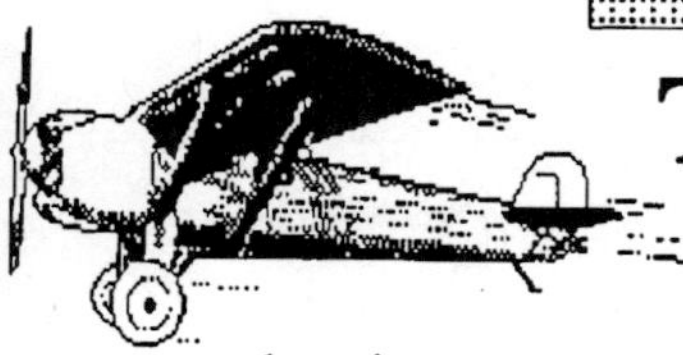

Tumbling made plain and simple tells you IN PLAIN LANGUAGE...How to load your machine, select the correct abrasives, compounds, tumbling speeds and intensities. How to use carriers, medias, fillers, additives, thickeners, suspension agents, pre-polishing agents. burnishing ...and the list goes on and on. Before you learn to fly...better learn how to ride. So lets start at the beginning...and if you are already an expert..then you can skip this part of the book.

---

**Anyone can learn how to tumble polish...if you follow the rules, keep a simple log of your tumbling operations, and remember..tumbling won't produce good results on bad materials. So were going to give you basic rules to follow: The THUMBS UP SIGN REPRESENTED BY "±" FOR THE "DO'S" AND THE THUMBS DOWN SIGN REPRESENTED BY "Ø" FOR THE DONT'S! [ Watch for these symbols throughout this entire book! ]**

---

Covered in the next few chapters are: basic types of tumbling equipment, how they differ in operation, locating and selecting materials, medias, carriers and fillers, abrasives, polishing compounds, and loading the tumbler correctly, for best results!

# SELECTING TUMBLING MATERIALS

## GEMSTONES

Wash and remove undesirable materials, let stand to dry.

SORTING

Ø FRACTURED---If fractures break the surface or appear to penetrate into the stones deeper than 2mm (about 1/16"), or are easily visible below the surface, sort materials out and save for later processing. (See CLEANING, bottom paragraph, page #42)

Ø PITTED---If pits appear deeper than 2mm (about 1/16") and/or narrower than 5mm, sort out these stones and save for later processing. (See Special Techniques, Treatments ,Tips)

Ø POROUS---Wet the cleaned materials using kitchen spray bottle filled with water and a few drops of detergent to break the surface tension. Sort out materials that soak up water quickly and the surface appears duller almost at once. These are probably too soft to polish well and could contaminate your load in later cycles.

***RULE # 1.***

Sort and process cleaned materials, saving pitted, and fractured materials for later treatment. Select more capacity than tumbling container. Load will be reduced 30 to 40% in grinding, requiring addition of more materials for best processing & tumbling action. (Covered in Loading, Grinding ) If excess is not available, select media to make up difference. (See sectiions #5-1 & 5-2 on Media) Good materials in...Good Materials out. The tumbling process cannot make good materials from poor but it will bring out the beauty...with a proper beginning.

# SELECTING TUMBLING MATERIALS

## GEMSTONES

Selecting tumbling materials by size is just as important as quality. Contact stone to stone, rolling action of the load, and in vibratory finishers both actions plus vibration serve to crush the abrasives and perform the grinding and polishing action. Sorting materials into small, medium, and large sizes, loading the barrel with a larger proportion of small in comparison to medium and large, results in the best tumbling action and results. Small 1/4" to 3/8", medium 1/2" to 1 1/2", large 1 3/4" and up.

---

*"Selecting materials by size...just as important as quality"*

---

Here's why size is important in processing (See illustrations) Loads with all large stones rotate unevenly which can result in longer processing times, chipping and spalling of the outer edges. This is especially true of softer materials.

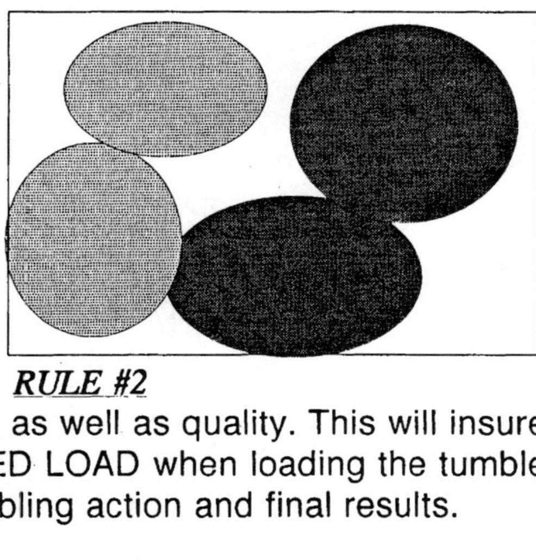

***RULE #2***

Sort materials by size as well as quality. This will insure selecting A BALANCED LOAD when loading the tumbler to promote the best tumbling action and final results.

# PLASTIC-CERAMIC-NATURAL-ORGANIC

Media, Fillers & Carriers - all similar in USAGE, but differ in description and materials, how to use, and in what types of equipment they are best utilized..

## PLASTIC

**PLASTIC**....Comprised of man made materials, the most common are plastic pellets of polyethelene & polypropolene ranging in sizes from approximately 1/8" to 1/4". These are best used in small rotary-barrel tumblers where soft to average materials are tumbled, or where large stones are in the load. Plastic media separates, help prevent fracturing, and where the load is very irregular in shape, improve the tumbling action. Plastic media are not dense or heavy, consequently do not add much weight to the load. Plastic Media are inexpensive and usually readily available from suppliers. Use from 20 to 30% Plastic Media by volume.

---

*"Medias, Carriers & Fillers...Similar in use...but differ in types, sizes, composition,...help prevent chipping"*

---

## CERAMIC

**CERAMIC**...A manufactured type material, available in many sizes and shapes, contains no abrasive. Angled cylinders 3/16" to 1/4" diameter, by about 1/2" in length are preferred, since contact with other materials is about 100% greater than with the similar sized round media. Load density and contact are key factors enhancing the tumbling process greatly. Ceramic cylinders serve as "roller bearings", improving tumbling action. Angled ends of the media serves as a Carrier, moving the abrasive grain into depressions and irregular areas, improving tumbling action,

# PLASTIC-CERAMIC-NATURAL-ORGANIC

& reducing tumbling times. Best used in vibratory Tumbling machines and medium Rotary Barrel tumblers. Use from 25% to 50% Ceramic media depending on the hardness, type and shape of materials being processed. Soft to medium hardness, larger, and irregular shaped loads, require more Media to keep the tumbling action smooth and even. Large or irregular shaped objects, flats, slabs, and pre-forms, should be surrounded with 50% or more media to carry the abrasives evenly to all areas.

## NATURAL

NATURAL media are usually similar in size but differ in shape, composed of stone such as Chert, Flint, Quartz. They can also be of the same materials as being processed, as long as the shape is rounded, not irregular and size is from 1/4" to 3/8".. Garnet is an excellent natural Media and Carrier, if available in correct sizes and shapes. It is dense, hard, adds considerable weight to the load, and when the load is completed, usable along with the rest of the finished materials.

## ORGANIC

ORGANIC...Natural, pre-shaped media such as wood. Available in several shapes and sizes including, cubes, diamonds, pointed pegs, or flat, to mention just a few. Especially effective when used in loads containing soft, dual hardness and tough materials, or materials that tend to undercut or "orange peel" when processing. These include Amber, Apache Tears, Azurite, Corals, Chrysocolla, Jade, Malachite,Turquoise, & Rhodocrosite. (These & other difficult materials will be covered in more detail under sections 32-1 & 32-2.)

# SILICON CARBIDE--BORON CARBIDE

## SILICON

**Silicon Carbide** abrasive grains (commonly called grit) are graded in numbered sizes from #36, #80, #60-90, # 60-120, #120, #220,F, #320, 2F, 400, #4-F, #600, #800,& #1000. The sm aller the number the larger the grain size resembling coarse sand for #36 down to a very fine flour for 1000 grit. Silicon Carbide has a unique property, when crushed or split, it always fractures with sharp points and edges exposed. Silicon Carbide is very effective in grinding almost any material, except Corundum, owing to its hardness of slightly over 9 on the MOH scale.

---

*"Silicon Carbide....always fractures with sharp points and edges "*

---

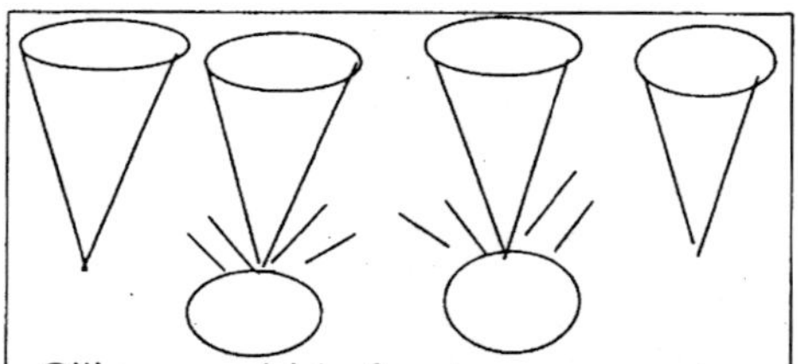

Silicon carbide fractures exposing sharp points. Over 2000s lbs of tumbling force can be applied.

*Grit* is used in the 1st & 2nd tumbling cycles (commonly referred to as grinding and sanding cycles) The roughness, shape, size & hardness, of materials determine grit size recommended. The more irregular, harder and larger the materials being processed, the coarser the abrasive grain size. Very Coarse grain size #36 is best used in large tumblers. The average grain size for (Grinding) is #80 or #60/90. (Sanding) #220 or #320.

| | |
|---|---|
| ± ± ± ± ± | Gem Materials 4+ to 8+ MOH |
| ± ± ± | Organics, Shells |
| Ø | Metals, Non-Ferrous |

# SILICON CARBIDE--BORON CARBIDE

## BORON

Boron Carbide is manufactured similar to silicon carbide, sometimes refereed to as Norbide® which is a trade name of the Norton brand. Grain sizes are available similar to silicon carbide except fewer grades are normally carried by suppliers. Boron carbide fractures similar to silicon carbide in usage and has one additional very important feature: Boron Carbide and Boron Nitride are two of the hardest substances known today, 9+ on the MOH scale.

Extremely effective for processing materials 9 in hardness which includes Corundum (Sapphire and Ruby---See Special Techniques and Treatments). Boron Carbide is very expensive, compared to Silicon Carbide, [up to 10 times] and not readily available. The other negative feature is a processing one; it is very, very, messy to use.

A special purpose in processing extremely hard materials, pre-forms, cabochons, or carvings from corundum should be a pre-requisite to consider using Boron Carbide, and the end results justify the high cost and special techniques. Vibratory finishers, especially those operating in the vertical plane provide adequate power to effectively use Boron Carbide. Since most corundum material is expensive, experimenting should be performed with caution, especially with the lower grades of Ruby Corundum.

| | |
|---|---|
| ± ± ± ± ± | SAPPHIRE CORUNDUM |
| ± ± | RUBY CORUNDUM |
| Ø | OTHER GEM MATERIALS |

# GRIT SIZE--MESH SIZE--MICRON SIZE

## ALUMINUM OXIDE

Aluminum Oxides are manufactured grit sizes corresponding to Silicon carbides, except much finer sizes are available up to 100,000 mesh (or Grit). The descriptive terms of Grit size, Mesh Size, and Micron Size tend to be somewhat confusing, describing particle size in different manners. Grit size is the physical size of the abrasive grain, or how large it appears to you when you look at it. Mesh size, actually refers to a screen through which various limits in particle size will pass during the grading process, which, in turn, governs the grit size. Micron Size, refers to the particle size of the abrasive grains in metric units. In the end this is a descriptive way of defining with different terminology a particular size of abrasive for use in processing.

"EXAMPLE.... 14,000 GRIT---EQUALS 14,000 MESH EQUALS ONE MICRON"

[ONE MICRON = 1/1000 OF A MILLIMETER. ONE MILLIMETER = 1/4000 OR 1/25th AN INCH. ONE MICRON IS ABOUT 40 MILLIONTHS OF AN INCH.

Micron sizes are handy in research measurements for surface analysis and RMS comparisons, the average person shouldn't be concerned. Grit sizes are most common up to 1000, after that mesh sizes are the most commonly used terms .

(COMPLETE ABRASIVE COMPARISON CHART IN LATER SECTION FOR THOSE WHO WANT TO GET INTO THE REAL "KNITTY GRITTY", HOWEVER!!

# SANDING--PRE-FINISHING--POLISHING

## ALUMINUM OXIDE

Aluminum Oxide abrasives are used primarily in the polishing process, but do find uses in the sanding and pre-polish of gem materials, organics, shells, and metals. Aluminum oxides are very hard, and vibratory processors are better suited for using these types of abrasives than rotary barrel tumblers. The reason for this is, they crush abrasives during the cycles, and aluminum oxide tends fracture with rounded rather than sharp points.

---

*"they crush...without producing sharp edges and points... as happens with silicon carbide"*

---

Some tumblers have a unique tumbling action of alternating separation and compaction which is accomplished during operation of the vibrating cycles leading the manufacturer to recommend that only Aluminum Oxide compounds be used in final polishing most gem materials. This is not true of all vibrating machines however, and other compounds are suitable for use.

Coarser sizes of Aluminum Oxides are used in the sanding and pre-finishing of softer and difficult to finish gem materials, organics, shells and non-ferrous metals. More detailed applications using aluminum oxides will be covered in sections 7-1, 25-1, 25-2, 25-3, 25-4, & 27-4.

± ± ± ± ± Pre-Polish, Polish Gem Material
± ± ± ± Pre-Polish Organics, Shells
± ± ± ± Pre-Finish Non-Ferrous Metals

# POLISHING COMPOUNDS--WET & DRY

## OXIDES

There seems to be no end to polishing-compound variety, at least that's the way it seems in viewing suppliers advertising, literature, instruction manuals, and catalogs. Most compounds used in polishing fall into the category of some form of oxide, normally used wet.

Oxides can be produced from refining and pulverizing natural earth elements such as tin, iron, chromium, and combining them with oxygen. When graded through fine mesh screens, usually 14,000 up to 100,000 mesh [1.0--.01 micron] they become suitable for use in polishing operations when added to water in tumbling containers.

---

*"synthetic aluminum oxides...recommended for vibratory finishers...natural oxides for rotary tumblers"*

---

Compounds are also produced synthetically using aluminum & zirconium, to form the oxides sold by many suppliers under a wide variety of trade names. Aluminum oxides are the most popular because of their lower cost and selection of mesh sizes available. Aluminum oxides are highly recommended for use in Vibratory tumblers, natural compounds, tin oxide, cerium oxide, iron oxide, and chromium oxide, are better used in rotary barrel tumblers. Synthetic compounds tend to be more economical.

The gentler action of rotary barrel tumblers does not "break down" the compounds as fast as the crushing action of vibratory finishers. Consequently natural oxides, which tend to be softer than synthetics, are more suitable for rotary processing in most small size machines with capacities from 1 1/2 to 12 pounds.

# POLISHING COMPOUNDS--WET & DRY

## DRY TYPES

Dry compounds were developed primarily for use in the pre-finishing and polishing cycles. They are compiosed of a carrier [usually some form of organic material] in varying grain sizes from #8 to approx. #36, which has been impregnated with a lubricant and abrasive grain for use in pre-finishing cycles, or a compound when designed for polishing operations. This type of compound is very effective when used in vibratory tumblers, not highly recommended for use in rotary tumblers. There's a reason for this! Rotary tumblers move loads very gently and slowly, and do not produce the intense crushing action of vibratory finishers. Most vibratory tumblers produce "squeezing" vibrations [1500 to 3600 per minute], in addition to rotary action, that is required to effectively use compounds that are embedded in other materials, not free rolling.

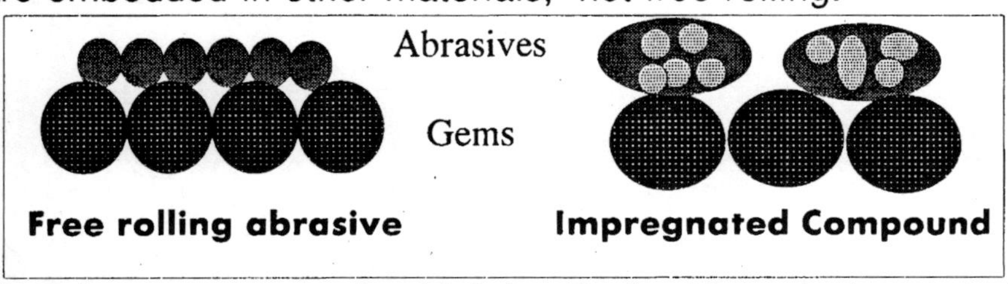


Abrasive or compound which is attached to or impregnated in another material, is less aggressive than if it were free to roll and expose new surfaces each time of rotation. {see illustration] The trade off is that the carrier medium effectively surrounds materials being processed with contact increased many fold. Materials of different hardness can be effectively pre-finished and polished in the same load, and dry compounds are re-usable. Chipping, spalling and fracturing are virtually eliminated.

± ± ±±± Vibratory–Gemstones, Organics, Shells, Metals
± ± Rotary–Metals

# BALANCED TUMBLER LOADS

## ROTARY

Correctly balancing gemstone, organic, shell, or metal materials for processing is one of the most important steps in tumbling process, yet the significance of this is often overlooked in the haste of getting started. Perform this step correctly, and make it your initial one, each and every time you process. It will reduce your tumbling problems, processing times can be reduced, and finished results will be vastly improved. Here's why...plain and simple.

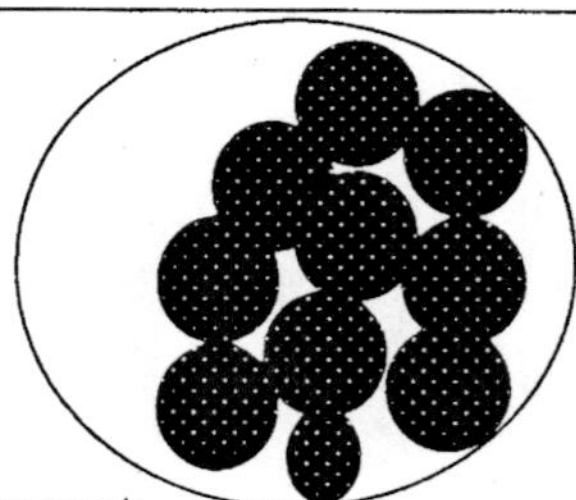

Incorrect

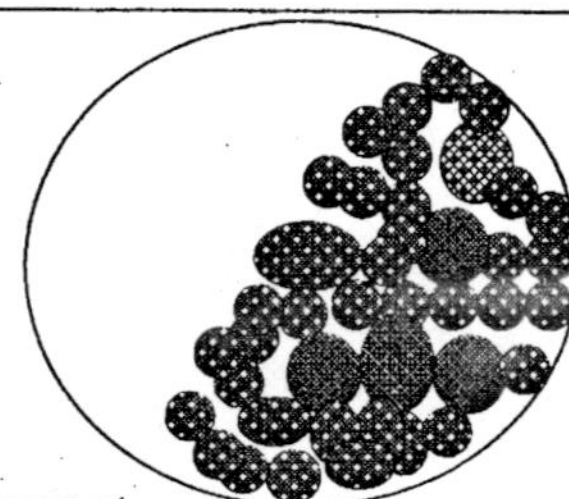

Correct

Rotary or barrel tumblers have a roll area and a slide area... do not load with materials which are all basically the same size, or all hammer broken... slide action is reduced, tumbling action poor. Note the same load with 25% smaller materials, balanced to fill in air spaces on rounded items, separate irregular shapes...improved rolling within the load, and better tumbling of materials. More contact in a rolling plane, rather than in a sliding plane. Fracturing, chipping, spalling reduced.......*abrasive contact improved*...shorter tumbling cycles, smoother action, uniform surfaces, and in the end...improved finished materials. What happened to the ***Airplane Special described*** in the introduction to tumbling...lets find out!

# BALANCED TUMBLER LOADS

## ROTARY

The airplane tire tumbler operation, was a "turn 'er on and let 'er go" type of start. As you can probably guess, the error was in not correctly sorting and balancing the load. (The large Lake Superior Agates in the 1 ½ to 3" size, plus abrasive, 1 pound to each 10 lbs of materials, water to cover, process for 1 week) The partially inflated tube confined the splash, however in gaining this advantage, the noise of the large agate stones IMPINGING in the tumbling area of the load went undetected. Even after abrasive slurries begin to thicken, a load of all large size materials can still be damaged, especially on the points and edges. The Lake Superior Agates had---fractured edges, spalling, and chipping---***there is a way to avoid this disaster.***

---

*"...add smaller materials...fill in spaces...between larger and irregular shapes. Balance the load,"*

---

Tumbler loads, whether gem materials, organics, shells, or metals, should not consist of all one size or shape. Carefully picture the load inside the container...add smaller materials to fill in spaces that exist between larger, and irregular shapes. Balance the load, not all large, not all irregular, not all small....but a combination. If varied sizes are not available, purchase a good media (such as ceramic angled cylinders), adding 25% to 50% media. Larger, harder, irregular shapes may require up to 50% media---smaller, smoother surfaces, require less. Softer materials require uniform, constant tumbling action. If you can detect uneven turning, the contact between the outer edges of your materials is "taking a beating", stop processing, ***balance the load and consider thickening.***

# BALANCED TUMBLER LOADS

## VIBRATORY

Balanced loading of vibratory finishers is just as essential, with one fundamental difference. The load vibrates in addition to rotating. So! So...the vibration causes materials to configure the load into a smaller, and more compact mass. Primarily, this takes place in the initial grinding stage, but will continue to some degree throughout the tumbling process. Here it is...*plain & simple*—when the load consists primarily of irregular shapes, or hammer broken materials, the entire contents of the container will rotate as one compacted mass, with very little rotation of the individual pieces, or in *Shakespearean*......."full of sound but signifying nothing". Very little abrasive reduction and or smoothing of the load will take place. Take extra precaution with vibratory tumbler loading.

---

*"Balance loading forstalls...One compacted mass...Little rotation of individual pieces...Take extra precaution with vibratory...loading"*

---

Ø....**Do Not** load the container with an entire load of hammer broken gemstone materials.

±....Do separate materials using small, rounded, smooth, media, 3/16" to 5/16" diameter using 25% for small to medium sized load up to 50% for large irregular shapes.

Ø....Do not use rough hammer broken or crushed material with rough surface, as media. (If you don't have natural material, consider purchase of ceramic media, 3/16" x 1/2" with angled ends).

Ø....Do not tumble all slabs or slices without 50% separation with media. Tumbling action will be uneven and slabs will lap together receiving virtually no treatment.

# BALANCED TUMBLER LOADS

## VIBRATORY

The contents of vibratory finishers operating in the vertical plane present a circular pattern of rotation around a center of the open container. There is no center projection around which the pattern of rotation forms. Horizontal plane units turbulate materials towards the center post as well as in a circle of movement around the post, similar to donught shaped load, vibrating as well as tumbling up and over. Balanced loading horizontal (donught or pumpkin shaped) tumblers is not quite as critical as vertical units. Care should be taken, with broken, irregular, and large materials, to use a mixture of sizes and media.

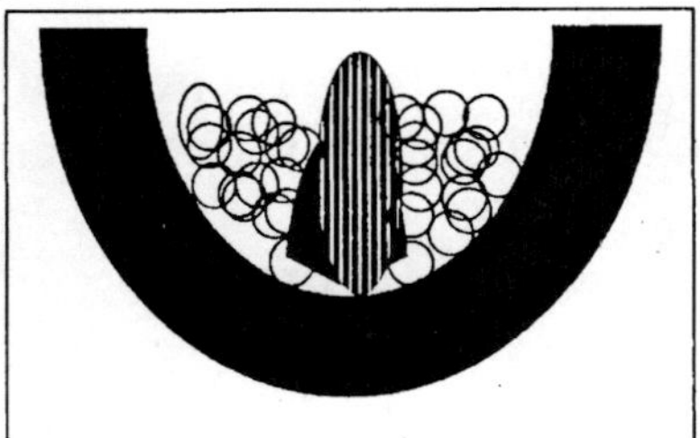

Horizontal type Machine

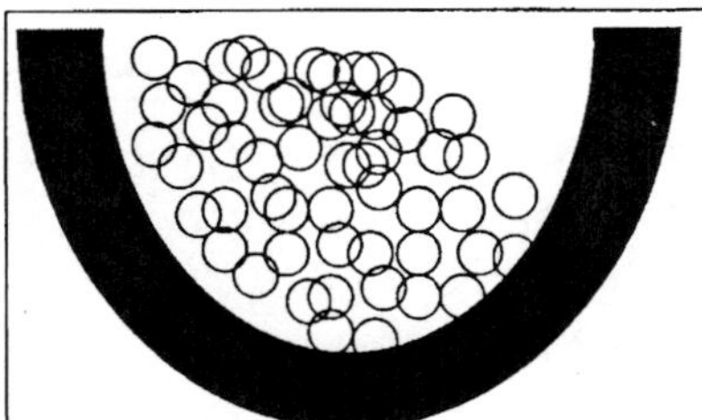

Vertical type Machine

Owing mainly to the horizontal spread, (breadth rather than height) for any give size load, natural force on each object is reduced. Spin or turn of objects enhanced, radius blending & rounding of edges & high points greater in this configuration. Shapes and dimensions however are not maintained as well as with vertical (non-center post) units. There are pros and cons for each type.

Spin off types, from the metal finishing industry, doughnut and pumpkin shaped units were not invented with gem tumbling in mind, rather for de-burring & radiusing metal parts and pieces.

# GEM MATERIALS-VARIETY-HARDNESS

## ROTARY

What about processing loads with a variety, different types and hardness of gem materials? While most manufacturers recommend loads of near the same hardness, it is possible to process and successfully finish mixed materials. True, it requires a little more planning, and care through the process, so if your one who abhors details, stick with the middle of the road. (quartz family, 6-7½ hardness, ½--1" size) If you are game to try here it is, ***plain & simple!***

Load carefully, no large materials (above 1" size), 30% to 50%, smaller 1/4" to 3/8" media size, rest of the load 1/2" to 1" mixed sizes. Nothing below 5 in hardness.

±....Treat and process the load, **for the softest material included.** If this is glass, or apache tears then process as if the entire load were this material.

±....Use finer abrasive for grinding stage (120-180)

±....Use suspension agent to thicken the slurry and cushion load. If not available use small plastic or wood filler material.

Ø....Large, flat slices or slabs in mixed load, may cause uneven processing, slide, roll, and in turn fracturing.

The physical size of softer materials is reduced rapidly. Softer smoother surfaced materials may be best initially loaded in the sanding stage, since the load level will have been reduced following the 1st Grinding Step, and will require additional material be added at this time to maintain a level for best action, maximum slide area, and optimum results. Use of suspension (thickening) agent should be considered, when processing mixed or full loads of softer smooth materials, 5 or below in hardness.

# FOSSIL IVORY-SHELL-CORAL-BONE

## VIBRATORY

Loading materials that are 3 1/2 to 4 on MOH'S scale requires that your media to load ratio be increased. A ratio of media to load of 1 to 1 up to 2 to 1 is not uncommon depending on the shapes of the materials. Processing requires maximum contact, even gentle tumbling action at reduced power or amplitude of tumbling intensity. Maximum contact between media and load is the key to success, and the load must rotate evenly.

Tumbling in a tightly knit mass, but still free to rotate gently within the load itself, the media must not only serve in effect as "bearings", promoting the smooth even tumbling action, the media must also provide maximum contact between the abrasive particles and the load, carrying abrasive slurry evenly to all parts of each individual piece contained within the load.

Levels in general should be in the 75-80% full by volume range. Vertical units need a small "pocket" near the top of the load to permit materials within the load to circulate. Loading too full or near the top will cause loads to rotate as one large mass, with little action within the load, thus very reduced processing results. Vibratory finishers are far more forgiving when it comes processing with loads which are not sufficient for the size of the container. Do not however go below the manufacturer's recommended minimum quantity for any given capacity of container or hopper. This can cause "jumping or hopping" of the load and damage.

Medias range from Ceramic and Hardwood to natural crushed shell. (Walnut, Pecan, Apricot, and Corn Cob)

# GEM MATERIALS-VARITEY-HARDNESS

## VIBRATORY

Loading materials of mixed variety and hardness for processing in Vibratory Finishers can be accomplished in somewhat the same manner as for Rotary Barrel tumblers. As stated previously, vibratory finishers are far more forgiving with improper loads than Barrel units.

One reason is that Vibratory units do not depend on maximum slide and roll area for effective processing. (See illustrations) Secondly the rolling or tumbling action does not primarily take place in a long slide area. Rather the effectiveness is dependent on high frequency vibration of stones within the abrasive slurry, rotation of individual pieces gently taking place within the mass of the load and not in a ***free slide area.*** For these reasons, mixed materials and mixed hardness [in the range of 5 to 7+] process nearly as effectively as if hardness or variety were primarily the same.

Tumbling action takes place evenly in the contact area. Vibration will provide the primary force within the load, with seconday roll action.

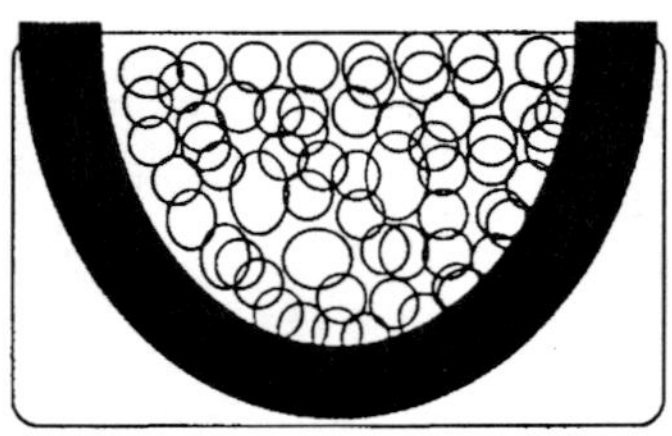

This is not "gospel" for all materials, but it holds true for the bulk of variety. It may appear that excess of instruction regarding loading has been made. Possibly like the fisherman buying the new lure, which produces nothing at his favorite "run". Then the foreign exchange student with jeans & sneakers, pulls 6 prize rainbows from the same waters, with the same lure. Must be the lure... couldn't be the presentation...our could it. Its important...or the amount of space would not have been designated.

# MIXING-VARIETY-HARDNESS

## VIBRATORY

Harder materials do not always polish to a higher degree, in any given period of time. Some, such as spinel, which is harder than topaz,. or corundum, which is harder than spinel, or jade, which is harder than glass, require that you consider this factor when loading for mixed processing.

Ø....Do not load too broad a range in hardness, because longer processing times for materials above 7+ in hardness tend to reduce the size of softer materials of 5 & 6 about 1/3 more than harder stones. ±....A range of 5-7 MOH'S scale will process with balanced loading. (Apache Tears to Agates) Media ratio 1/4 to 1/3 by volume, remainder of the load in materials in the 1/2" size up to 1" size, to provide even rotation and maximum contact area between materials within the load. No more than 1 or 2 larger stones above the 1½" size, per 5 lbs. of total load, otherwise fracturing & spalling will occur, especially on the larger pieces.

Gem Material "Toughness" scale rates materials for their tenacity, with some surprising incongruties. Diamond is natures hardest known material, but not for toughness, in fact it can be crushed or broken much easier than Jade. *As a general rule: harder materials will require more time in Grinding and Sanding cycles, and less in Polishing cycles, with just the opposite for softer-tougher materials* Here is one of the least understood fundamental principles of mass finishing to be covered in "Final Polishing". Why do the harder materials, rounder surfaces, and edges seem to have polishing priority? Learn exactly why!

# SORTING-MEDIAS-ABRASIVES-POLISH

## GEMSTONES

- SORT..Cull fractured, pitted, porous & irregular shapes. [Save for later use]
- SELECT..Grade for sizes [1/4"-3/8"]....[1\2"-1 1/2"]...[1 3/4"-UP]
- MEDIA Plastic-Ceramic-Natural-Organic. Fill in spaces, Soft..Large..Irregular shapes. Reduce chipping.

| | Abrasive | Hardness | Material |
|---|---|---|---|
| • ABRASIVES.. (Grinding & Sanding) | Aluminum Oxide.. | 2---4 + | Organics-Metals |
| | Silicon Carbide.... | 5------8 | Gem Materials |
| | Boron Carbide..... | 8 + --9 | Gem Materials |

| | Compound | Material | Tumbler |
|---|---|---|---|
| • PRE-POLISH | Aluminum Oxide #400-600-800 Wet slurry | Soft 2--6 + Gemstones, Organics, Metals | Rotary Vibratory |
| | Vibra-Grind® Dry | | Vibratory |
| • POLISHING | Natural Oxides: Cerium, Chrome, Iron, Tin, | Average 5----7 + Gemstones | Rotary |
| • POLISHING | Synthetic Oxides 1/4 to 1 micron (14,000-100,000 Aluminum Oxide) | Gemstones Average---- Hard | Vibratory |
| • POLISHING | Wet'n-Dry™ | Gems-Metals, | Vibratory |

# LOADING-GEMS--ORGANICS--METALS

## ROTARY

| | |
|---|---|
| • VOLUME | 25--50% Smaller materials or Media<br>Load 50-60% full by Volume for maximum effect in upper slide area. Higher volume media with soft, irregular, large materials. |
| • HARDNESS | 5---7 + HARDNESS mix 1/2--1" Size<br>Process large flat slabs Seperately<br>Process Mixed Gem Materials for softest<br>Thicken or Cushion for loads below 7<br>Higher volume of media for soft materials. |
| • MIXED GEMS | Soft 4--5 wears down 1/3 more than hard.<br>Load in Sanding Stage or use larger sizes.<br>No more than one or two 1½" per 5 lbs. |
| • MEDIA | Plastic 1/4" "Poly" pellets use in----Rotary<br>Natural materials 1/4 to 3/8" size------Both<br>Synthetic Ceramic cylinders 3/16th---Both<br>25%--50% by volume based on materials |
| • METALS | Finishing, Burnishing, Polishing Metals.<br>Load levels--maintain 60% barrel capacity<br>Media---3 to 6 parts media to 1 part load. |
| • MEDIA | Plastic (Abrasive charged Triangles-Cones)<br>Ceramic Angled Cylinders 1/16--1/8" dia.<br>Hardwood cubes, pointed rectangle 1/8-¼"<br>Steel Balls, Pins, Balcones, 1'/16"--1/8"<br>(Burnishing with steel requires use of special burnishing soap & water solution) |

# LOADING-GEMS-ORGANICS-METALS

## VIBRATORY

- **VOLUME** — 25-50% Smaller materials or media Load 75%-80% full by volume for best rotation. Higher volume of media with soft, irregular, and large materials. Do not load all hammer broken or irregular.

- **HARDNESS** — 5--8 hardness mixed sizes 1/2"---1"
  Flat slabs over 2" best results separate
  Process for softest contained in the load
  Sanding stage best for staring soft stones
  Higher ratio of media to stones for soft

- **MIXED GEMS** — Soft 2½-4½ best processed separate
  Soft 5-6 best results..start in sanding step
  Soft 5-6 wear down 1/3 more than hard
  Limit large stones of one or two per 5 lbs
  Organics----Fossil, Shell, Bone separately

- **MEDIAS** — Natural 1/4" to 3/8" 1/4 to 1/3 by volume
  Synthetic ceramic cylinders 3/16" angled
  Too much media below 1/4" slows results

- **METALS** — Load levels---75%--80% full by volume
  Media---3 to 5 parts media to 1 part load

- **MEDIAS** — Poly, polished ceramic, hardwood add to heavy loads to attain best volume level, maximum rotation and action. Steel balls, pins, ball-cones 1/16--1/8" require special soap & water solution. Organics--Wood, cob, shell (Apricot, Pecan, Walnut) dry.

# ORGANICS [ BONE-CORAL-SHELL ]

## VIBRATORY

| | |
|---|---|
| • VOLUME-LOADING | 75% TO 80% full by volume with media ratios up to 2-1 depending on shapes-sizes Flat pieces----Separate--use special medias |
| • CONTACT | Load-Media---Smooth rotation, full contact Carry abrasives---uniformly to entire load Rotation within load essential, especially for irregular shapes. |
| • MEDIAS | Natural---1/4--3/8"--Pre-finishing wet<br>Ceramic---1/8"--3/16---Pre-Finishing wet<br>Organic--Hardwood shapes 1/8"--3/8" and<br>Organic--Apricot, Pecan, Walnut, CobPre-Charged hardwood sawdust, all effective in Pre-Polish & Polish Cycles |
| • MIXED Ø | Properties of most organics are unique. Separate materials for best results. (See sections 69-1 and 70-1) |

Organic materials ( Amber, Bone, Coral, Ivory, Pearl, Shell, Teeth) aside from being very soft [2-4 in hardness] have individual unique properties. Irregular shape, absorbent, some tough, some brittle, some flexible, some resilent---all in all special compounds, media, methods, techniques, processing, and polishes are normally employed for each particular type material. The anamolistic nature of organics due to species, type, location, age, porosity in addition to the above irregularities afore mentioned, make standards difficult to establish. One recognized processing criteria is evident---DRY FINISHING (Dry Pre-Polish--Dry Final Polish)

# TUMBLING DATA-RECORDS-NOTES

| DATE | MATERIALS | MEDIA | ABRASIVES | LIQUIDS | TIME |
|---|---|---|---|---|---|
| | | | | | |
| | | | | | |
| | | | | | |
| | | | | | |
| | | | | | |
| | | | | | |
| | | | | | |
| | | | | | |
| | | | | | |
| | | | | | |
| | | | | | |
| | | | | | |
| | | | | | |
| | | | | | |
| | | | | | |
| | | | | | |
| | | | | | |
| | | | | | |
| | | | | | |
| | | | | | |

# TUMBLING DATA-RECORDS-NOTES

| DATE | MATERIALS | MEDIA | ABRASIVES | LIQUIDS | TIME |
|---|---|---|---|---|---|
| | | | | | |
| | | | | | |
| | | | | | |
| | | | | | |
| | | | | | |
| | | | | | |
| | | | | | |
| | | | | | |
| | | | | | |
| | | | | | |
| | | | | | |
| | | | | | |
| | | | | | |
| | | | | | |
| | | | | | |
| | | | | | |
| | | | | | |
| | | | | | |
| | | | | | |
| | | | | | |

# REMOVE...REDUCE...RE-STRUCTURE

## STEP #1

In the quest for a "perfect polish" (which exists only in theory), enhance the lines, pattern, contrasts, color, symmetry, and beauty, one must begin by preparing the surface of the materials for polishing.

#1. Remove excess material improving the appearance

#2. Reducing the rough exterior to a smooth even texture

#3. Re-structure the surface to a uniform "scratch pattern"

DURING THE EXPLANATION PROCESS there seems an increased capability to retain knowledge, if one understands why you need to know in the first place. Just an explanation doesn't really place any values or set any priorities. Set forth on the following page are graphic illustrations of actual surfaces.

Due to the curved parabolic surfaces of most gem materials, a Taylor-Hobson surface measurement unit was used. Tracing the surface profile (viewing the surface of material ,as if it were sawn in half), showing the peaks and valleys. Data is then amplified from the analog (mechanical) readings to an accurate electronic image (which can be magnified 10--1000 times), accurately and visually on an oscilliscope screen. Recording measurements from rough initial surfaces, prior to any grinding...comparing at various stages during grinding, and when a uniform scratch pattern (surface) with no further improvement apparent. A visual conception of what takes place in the grinding process *REMOVAL...REDUCTION...RE-STRUCTURING!*

# SELECTING ABRASIVE SIZE

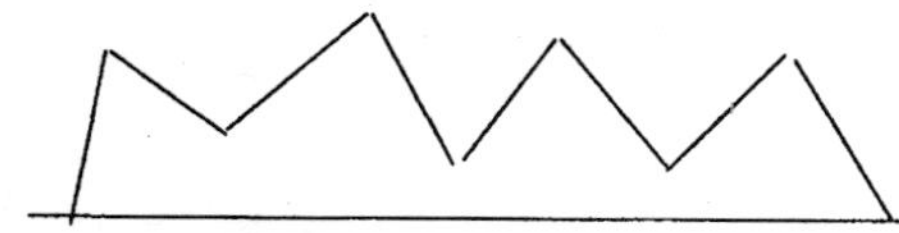

Typical broken surface

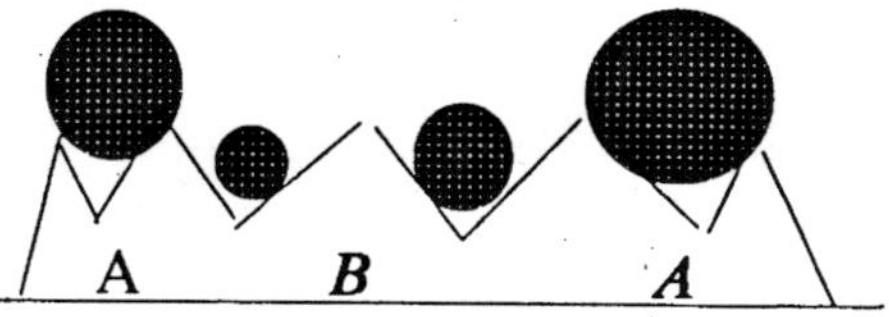


Abrasive Grit Size

The objective: Remove excess material, Reduce the height of the peaks, Re-structure the surface to a uniform pattern.

*(A)* .....Abrasive size large enough to span the peaks. will reduce the height of the peaks...even the texture.

*(B)* .....Abrasive size too small will texture the surface deeper, with little reduction in the height of the surface.
This is why tumbling progresses in cycles using selected sizes of abrasives depending on materials to be processed.

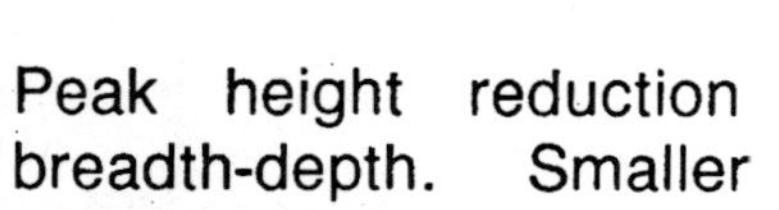

Surface after grinding

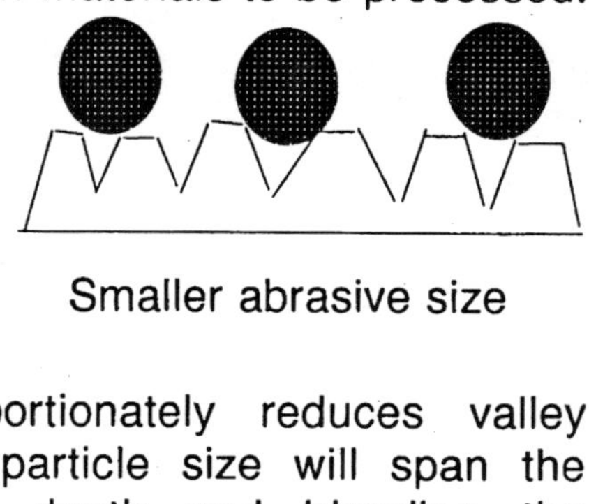

Smaller abrasive size

Peak height reduction proportionately reduces valley breadth-depth. Smaller grit particle size will span the valleys...further reducing the depth and blending the surface, during STEP #2 SANDING CYCLE.

Using incorrect abrasive..with very rough textured surfaces in STEP #1...can result in very slow processing. If the surface is not blended, smaller abrasive sizes used in subsequent cycles could INCREASE the depth of the valleys. The peaks must be reduced...texture blended.

# ABRASIVE GRAIN COMPARISON CHART

| | Hardness | | Properties | Shape |
|---|---|---|---|---|
| Alumina | 9 | 2,000 | Synthetic | Friable |
| Alum. Oxide | 9 | 2,100 | Synthetic | Wedges |
| Synth.Sapphire | 9 | 2,550 | Synthetic | Hex. |
| Silicon Carbide | 9+ | 2,550 | Synthetic | Wedges |
| Boron Carbide | 9+ | 2,880 | Synthetic | Slate |
| Diamond | 10 | 7,000 | Natural-Synthetic | |

Silicon Carbide

| Mesh | Micron | Recommended Use | (HARDNESS) |
|---|---|---|---|
| 80 | | Grinding | [7-8½ Mohs] |
| 100 | 150 | Grinding | [7-8½ Mohs] |
| 120 | 100 | Grinding | [6-7½ Mohs] |
| 220 | 70 | Sanding | [7-8½ Mohs] |
| 320 | 45 | Sanding | [6-7½ Mohs] |
| 500 | 30 | Sanding | [7-8½ Mohs] |
| 600 | 25 | Sanding | [6-7½ Mohs] |

| Aluminum Oxide | | Rotary -- Vibratory | |
|---|---|---|---|
| 400 | 35 | Pre-Polish | [6-7 Mohs] |
| 600 | 25 | Pre-Polish | [3-6 Mohs] |
| 14,000-- | 1 | Final Polish | [5-8 Mohs] |

| Wet'n Dry Compounds | | Vibratory Only | |
|---|---|---|---|
| 600 | 35 | Sanding | [7-8½ Mohs] |
| 800 | 20 | Sanding | [3-6 Mohs] |
| 3000 | 5 | Pre-Polish | [7-8½ Mohs] |
| 9000 | 3 | Pre-Polish | [3-5 Mohs] |
| 18000 | .75 | Final Polish | [7-8½ Mohs] |
| 60000 | 0.3 | Final Polish | [3-6 Mohs] |

| Polishing Oxides | Rotary |
|---|---|
| Cerium, Chrome, Iron, Tin (oxides) | [4-8 Mohs] |

# GEMSTONES-GEM MATERIALS

## ROTARY

| GRIT SIZE | MATERIALS-TEXTURE | HARDNESS |
|---|---|---|
| #60-90 | Rough Surface Texture | 7--8½ |
| #120-180 | Medium Surface Texture | 5-6½ |
| Silicon carbide | 2 to 2½Tablespoon's per pound of load or, 2 to 2½ oz. (by volume) per pound | |
| Load Capacity | 60--70% to capacity of barrel | |
| Water level | Just below final layer of materials | |
| Run time | 7--10 days (average processing time) | |
| Venting | After 1-2 days, relieve gas pressure | |

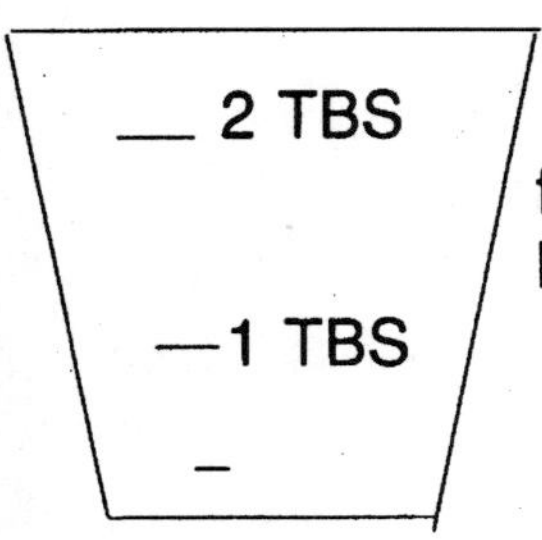


A handy way to measure abrasives- and compounds. Purchase a supply of plastic ,medicinal measuring cups from local medical supply house. These have marks-----Tablespoons, Ounces, ML. Accurate, Clean, Convient, Disposable.

(Sprinkle evenly over the load)--------Add a few drops of photo-flo®, (obtainable at most photo supply stores) baking soda, Sal-lSoda, or de-bubbler (used in investment casting processes)...Just enough to break the surface tension of the water. This helps to stop the foaming reaction experienced with some types of gem materials.

Make certain you have a balanced load for materials being processed. Add Thickening agent if the load contains materials 6 or below in hardness., especially when processing brittle materials, with poor toughness. Too much suspending liquid can retard action, and stock removal.

# GEMSTONES-GEM MATERIALS

## VIBRATORY

**GEMSTONES**

| GRIT SIZE | MATERIALS-TEXTURE | HARDNESS |
|---|---|---|
| 60-100- | Rough-Broken Surface | 7-8½ |
| 120-220 | Medium-Natural Surface | 4½-6½ |

Silicon Carbide: 1½-2 Tablespoons per pound of Load or by weight 1½-2 oz. per lb. of load

Load Capacity: 70%-85% to capacity of container

Water: Add amount equal to abrasives or rinse load, drain off all that will freely run. The remaining liquid, will be sufficient to start.

Abrasives: Sprinkle slowly over the load while unit is running, to distribute evenly through load.

Defoamer: A few drops of investment de-bubbling liquid can help reduce foaming sometimes experienced with some gem materials.

Cover: Use clear kitchen wrap, (multiple layers if required) stretch over container opening, using as a gasket to retain moisture. Place factory container cover over top.

Slurry: Vibratory units process with batter like mixtures..a consistency that will usually require additional liquid during processing. As materials are reduced by grinding, slurry will thicken. Some materials may require multiple grinding cycles, removing excess slurry-mud mix and repeating cycle.

# GEMSTONES-GEM MATERIALS

## VIBRATORY

SPECIAL TREATMENT---Materials in the hardness range from 4½-6...Smooth natural surface (water worn or wind worn) without pits or depressions. Examples: (Apache Tears (Glass) Chrysocolla, Coral, Feldspar, Hematite, Jade, Lapis, Malachite, Opal, Spodumene, Turquoise, Zoisite)

GRIT SIZE #220--320 1-2 tablespoons per lb. of load

WATER Use equal amount of water (Distilled or bottled recommended)

ADDITIVES Thickening-Suspending agent...add to load to cushion against chipping/spalling when processing materials with poor toughness or below 5 in hardness. (Available as a liquid concentrate, or popwder.)

RUN TIMES 2-4 Days..Check load--if slurry is thick and stops rotation, remove excess mud, wash and clean..re-start with new abrasive, if surface texture is not as desired. Repeat cycle if required to produce uniform texture.

---

***Soft, Smooth, Poor Toughness...use finer abrasives***

---

Processing smooth-surfaced, softer materials, or those with poor toughness, using abrasives that are too aggressive, or large in grit size, will produce a surface texture requiring extra processing to prepare for polishing. Surface pattern produced will be deeper in texture, and may not reduce uniformly in subsequent operations, sufficiently for polishing. Polish will appear hazy.

## GEM-MINERAL SCALE OF HARDNESS

The property of a mineral or gem material to resist scratching or abrasion is referred to as hardness. This differs from the penetration (ductile) hardness of metals. Both have comparative scales of measurment...*BRINELL* for metals, *MOH'S* for gem materials, generally referred to as the MOH scale of hardness.

| *GEM-MINERAL* | *MOH SCALE* |
|---|---|
| Diamond | 10 |
| Corundum | 9 |
| Topaz | 8 |
| Quartz | 7 |
| Feldspar (Orthoclase) | 6 |
| Apatite | 5 |
| Fluorite | 4 |
| Calcite | 3 |
| Gypsum | 2 |
| Talc | 1 |

Practical, and handy...test set made from common materials...Carry with you when collecting. Epoxy on a 1/4" or 3/16"hardwood dowel 3" in length. Each will scratch the material below but not above.

| | |
|---|---|
| *9* | *Chip of Ruby-Sapphire or Silicon Carbide* |
| *8* | *Chip of Topaz or Synthetic Spinel* |
| *7* | *Chip of Quartz Crystal* |
| *6* | *Chip of Feldspar (Tempered Wire or File Section)* |
| *5* | *Chip of Glass or section of knife blade* |
| *4* | *Chip of Fluorite (Fluorspar) or hard copper pipe* |
| *3* | *Piece of Soft Copper or copper penny* |
| *2-2½* | *Piece of Ivory ( Or use finger nail as test)* |

## GEM-MINERAL SCALE OF TOUGHNESS

The ability of a gem or mineral to resist an applied force to its rupture or breaking point, is a gauge of that materials relative toughness. Most scales of tougness list material's ranging from excellent to very poor. This scale lists Gems-Minerals on a scale of **1 to 10** as well as relative toughness.

| RATING | *GEM-MINERAL* | *SCALE* |
|---|---|---|
| EXCELLENT | Jade, (Nephrite), Jadite | 10 |
| VERY GOOD | Ruby, Sapphire, Alexandrite Diamond | 9 |
| GOOD | Amethyst, Jasper, Quartz, Citrene | 8 |
| GOOD-FAIR | Aquamarine, Garnet Tourmaline, Spinel, | 7 |
| GOOD-POOR | Coral, Turquoise, | 6 |
| FAIR | Ivory | 5 |
| FAIR-POOR | Moonstone, Lapis, Pearl | 4 |
| POOR | Amber, Emerald, Shell, Tanzanite | 3 |
| POOR-VERY POOR | Peridot, Rutile | 2 |
| VERY POOR | Amber, Opal | 1 |

# INTERMEDIATE TREATEMENT

## GEMSTONES

Sanding cycles 1 & 2 are critical tumbling processes! PRIOR TO processing in SANDING-1...WASH LOAD THOROUGHLY. Do not let the load stand motionless in barrel or hopper, (contents tend to cake or harden) it is best to clean immediately after completion of step #1 (Grinding). Rotary barrels can be opened carefully, (with lid in upright position), tap water added if contents are thick.

## ROTARY–VIBRATORY

A handy cleaning setup is 5 gallon pail, 2 plastic dishpans, and colander. Add 1" water to dishpan, place plastic colander in bottom and pour contents of barrel or hopper into colander. (Note: Rounded Collander with flat base works best) Flush water over load and rotate collander by hand to remove excess sludge & grit. remove the collander and contents...set aside in 2nd dishpan to drain.

Stir water-sludge mix of 1st container and immediately pour entire contents into 5 gallon bucket. Cover and set bucket aside to settle contents, for later treatment & disposal. (*DO NOT POUR INTO ANY DRAIN SYSTEM* Silicon Carbide\Water mix can set up like cement.

Thoroughly clean load removing all mud, sludge, and grit. Remove colander & load...set aside in clean dishpan to drain. Pour excess from dishpan into 5 gal. bucket to settle. Next day water will clear, contents can be poured off. Excess grit in bottom of bucket can be salvaged & re-used in another ***grinding*** cycle! Pour contents of colander into clean dishpan for inspection and sorting in preparation for the next step in processing...***SANDING***.

# INTERMEDIATE TREATMENT

## GEMSTONES

Prior to the next cycle in processing, sort deeply pitted & rough surface material for processing in a future grinding cycle. This would be stones that appear to have narrow pits of approx. 2-3mm in width & depth (about 1/8"--3/16") Surface reduction in the Sanding and Pre-finish may not reduce or remove sufficient material to texture the surface for polishing...especially on materials 7-8+ in hardness.

---

*"Its sometimes best to process through two Grinding Cycles"*

---

If your load is below 7 in hardness ( 5-6+hardness) or a mixed load...delay the sorting of softer materials until the **Sanding Cycles** are completed. These materials may experience more stock removal during processing, and blend the rough or pitted & rough areas.

Materials below 5 in hardness were probably not mixed for processing in the initial grinding cycle These materials are best started or added in the next step of **Sanding-1**, if the surface requires reduction to a smooth texture. Smooth surfaced, below 5, can be added or started in **Sanding-2.**

Materials with deep, visible fractures (especially those that break the surface) should be sorted out for re-run in future **Grinding** cycles. It is sometimes best to process this type of material through two **Grinding** cycles before proceeding to **Sanding-1**. This way you have sufficient materials to replace those sorted out, also sufficient stock to make up for the physical reduction in the volume of the load during *Reduction* in the **Grinding** cycle.

# GEMSTONES-GEM MATERIALS

## ROTARY

| GRIT SIZE | MATERIALS | HARDNESS |
|---|---|---|
| 180-220 | Agate, Beryl, Chrysoberyl, Topaz, & Similar materials | 7--8 1/2 |
| 2/F-320- | Amethyst, Apache Tears Feldspar, Garnet, Glass, Hematite, Jade, Lapis, Opal, Peridot, Turquoise, Zircon | 5--7 |
| Silicon Carbide | 2--2 1/2 Tablespoons or 2--2 1/2 oz per pound of load | |
| Load Capacity | 60-70% of Barrel Capacity. Add pre-tumbled materials or Ceramic Cylinders\ Balls as required.. bring load to capacity | |
| Water Level | Just below final layer of materials. | |
| Run Time | 7-10 days (average processing time) | |
| Venting | After 1-2 days relieve gas pressure | |

Add a few drops of Photo-Flo® or investment De-bubbler (or even diswasher gel) to break surface tension. Add thickening or suspending agent if load contains materials 6 or below in hardness, especially if the materials are brittle or of poor toughness. Do not thicken the load excessively as tumbling action can be retarded....this can cause excessive tumbling times, little stock removal, and poor surface blending.

# GEMSTONES-GEM MATERIALS

## VIBRATORY

| GRIT SIZE | MATERIALS | HARDNESS |
|---|---|---|
| 120\220 | Agate, Beryl, Chrysoberyl, Topaz, & Similar Materials | 7--8 1/2 |
| 2\F-220 | Amethyst, Apache Tears, Feldspar, Garnet, Glass, Hematite, Jade, Lapis, Opal Peridot, Turquoise, Zircon | 5------7 |
| 3\F-400 | Amber,Bone,Coral,Ivory, Nuts, Organics, Seeds, Shell | 2 1/2--4 1/2 |
| Silicon Carbide | 1 1/2--2 Tablespoons per pound of load, sprinkle evenly into load while unit is running | |
| Load Capacity | 70--80% to capacity of container | |
| Water | Add water equal to the amount of abrasives used | |
| Cover | Use clear kitchen wrap (multiple layers) stretch over container opening, making gasket to retain moisture...place factory cover over top to secure gasket in place | |
| Slurry | Maintain batter like consistency during processing...add water as required. | |
| Run Times | 2-4 days depending on materials. | |

# GEMSTONES-GEM MATERIALS

## TUMBLING TIPS: ROTARY

Sanding-2 is basically a duplicate of Sanding-1, but with several very important exceptions:

1. Abrasives grains used will be much finer in size.
2. Slurry mixes will be much thinner in consistency.
3. The last cycle prior to polish for average materials.

---

*"damage may not be detected until polish cycle is completed"*

---

Smaller abrasives grains will maintain more points of contact between materials being processed. Surface blending is at its optimum in this cycle. It is important that balanced loads (ratio of small and larger materials) are maintained. Load levels must be correct to maintain smooth and even tumbling action.

***Thinner consistency in slurry mixes will result in freer movement of the load consequently...damage by contact between materials can occur in this cycle. This damage may not be detected until the polish cycle is completed!***

For most materials of 7--8½ in hardness, this will be the last cycle before polishing, consequently, it is the last chance to properly prepare materials for the final polish. Anyone who has been involved in the lapidary process will verify...good stones are "***made* in the sanding process**". Incorrect processing in this cycle can result in disaster in the final polish cycle. Make no mistake, this is one of the most important...if not the most important tumbling cycle!

# GEMSTONES-GEM MATERIALS

## TUMBLING TIPS: ROTARY

So, now that we have your attention, lets see if there are some tumbling tips that may help to avoid disaster, especially on softer materials and those of poor to very poor toughness. [Hardness range from 5--6½...Toughness range from 5-6.]

Tip #1 Load should not contain materials below 5 in hardness.

Tip #2 Load that contains mixed materials...hardness below 7...toughness fair--very poor...process for the softest material contained in the load.

Tip #3 Cushion the tumbling action by using various methods for thickening the slurry mixture...any of the following have been used effectively by the author: Syrup, Sugar, Molasses, Rice Hulls, Gelatin,Corn Starch, Sawdust...and the list goes on. The principle is to buffer or control impacts.

Tip #4 Don't cushion to the point of retarding action so that practically no results occur. Listen to the sound of the load...usually you can tell when the action has an even, smooth sound.

Tip #5 Propylene Glycol (R.V. Antifreeze) can be used (with water if required), to provide cushioning!

## GEMSTONES-GEM MATERIALS

### TUMBLING TIPS- VIBRATORY

The same basic exceptions apply to vibratory finishing. As finer abrasive sizes are used, slurry mixes become much thinner, and for most materials, this may be the last stage prior to the final polish.

---

*"vibratory tumblers are more forgiving than rotary machines"*

---

Abrasive grains are only about 1/2 the size of those used in Sanding-1. A much tighter abrasive pack will occur in this stage with contact points at least quadrupled. Blending and smoothing of the surfaces will be at the highest levels of any of the pre-finishing cycles, during Sanding-2. This requires that load levels and load balance (ratios of smaller to larger materials) be correct in order to maintain smooth and even tumbling action, even with vibratory tumblers.

Vibratory finishers will have thinner slurry mixes and damage to loads of mixed hardness containing materials below 6 can occur. The possibility, however, is not nearly as great as with rotary units. Vibratory loads pack tighter with only the load rotating, and the container practically motionless, the possibility of damage is minimal.

While thickening of the slurries in Sanding-2 is certainly a recommendation...especially if materials 5 or below in hardness are contained in the load..slurry consistency is not nearly as critical as with rotary-barrel units. This cycle is still one of the most important in tumbling as for most

# GEMSTONES-GEM MATERIALS

## TUMBLING TIPS-VIBRATORY

Materials from 5-8 in hardness can be processed in the same load...special attention to balanced ratios of small to large sizes should be observed. Remember..the slurry mix is thinner and the larger the material size...the easier it is to have damage occur...especially to softer materials. (5 and lower in hardness)

Tip #1 Vibratory finishers process mixed loads with a minimum risk of sub-surface fracture or spalling of most materials 6--8 in hardness.

Tip #2 Slurry mix should be thickened if load contains materials below 5 in hardness--toughness.

Tip #3 Vibratory tumblers with variable speed control virtually eliminate the need for additives and or thickeners. Exception-materials 4 Moh& below.

Tip #4 Start the cycle on medium setting with tumblers having variable speed control. This will give the finer size slurry mix, greater ability to blend the "scratch pattern"from the previous step. Increase the intensity midway in the cycle...this will "break down" the abrasive during the last part of the sequence. The end result will be finishes that are easier to polish...and higher reflective finishes will result.

Tip #5 Thickening\Cushioning agents are available from suppliers...or use Tumble-troll C®.

# GEMSTONES-GEM MATERIALS

## ROTARY

| GRIT SIZE | MATERIALS | HARDNESS |
|---|---|---|
| 5/F--500 | Agate, Beryl, Chrysoberyl, Topaz, & Similar materials | 7--8 1/2 |
| 600 | Amethyst, Apache Tears, Feldspar, Garnet, Glass, Hematite, Jade, Jasper, Lapis, Opal, Peridot, Turquoise, Zircon | 5--7 |

Silicon Carbide: 2--2 1/2 Tablespoons or 2--2 1/2 oz. per lb. of load.

Load Capacity: 60-70% of Barrel Capacity. Add pre-tumbled materials or Ceramic Cylinders\Balls as required...bring load to correct level.

Water Level: Just below final layer of materials.

Run Time: 5--7 days (average processing time)

Venting: After 1--2 days relieve gas pressure.

Adding a few drops of non-suding soap or investment de-bubbler to break the surface tension of the water, may enhance the tumbling action. Add thickening or suspending agent if load contains materials 6 or below in hardness..or poor toughness. Do not thicken excessively, the objective is to cushion the load, preventing spalling, not to retard or stop proper tumbling action of load within the barrel.

# GEMSTONES-GEM MATERIALS

## VIBRATORY

| GRIT SIZE | MATERIALS | HARDNESS |
|---|---|---|
| 4\F--500 | Agate, Beryl, Chrysoberyl, Topaz, & Similar Materials | 7--8 1/2 |
| 600 | Amethyst, Apache Tears, Feldspar, Garnet, Glass, Hematite, Jade, Lapis, Opal, Peridot, Turquoise, Zircon | 5--7 |
| 600--800 | Amber,Bone, Coral, Ivory, Organics, Seeds, Shell | 2 1/2--4 1/2 |
| Silicon Carbide | 1 1/2--2 Tablespoons or 1 1/2--2 oz. per pound of load...sprinkle evenly while unit is running | |
| Load Capacity | 70--80% of capacity of container | |
| Water | Add water equal to the amount of abrasives used. | |
| Cover | Use clear kitchen wrap (multiple layers) Stretch over container opening to retain moisture....place factory cover over top. | |
| Slurry | Maintain batter like consistency during processing...add water as required. | |
| Run Times | 2--4 days depending on materials. Brisk rubbing of sample on clean felt or wool cloth should create dull gloss...3-4 days. | |

# BEFORE PRE-POLISH--POLISH CYCLES

## GEMSTONES

"Cleanliness is next to Godliness"...a truer statement was never made, especially when it comes to "mass finishing" or tumble processing. The advantage of tumbling is that you're finishing a quantity of materials (over a longer period of time) rather than hand finishing 1 piece (over a shorter period of time). There is less labor involved, and you can inspect the progress, make processing corrections if required, and produce highly polished materials.

---

*"the importance of making certain that no abrasives or residue are carried over to the next cycle cannot be stressed enough"*

---

This is "tumbling in a nut shell"....do it right and the end results will be as you anticipated. As the old maid (who always wore seven-button shoes, mainly because she couldn't seem to "fasteneight") once said..."had I known the realization were as great as the anticipation, there never would have been any hesitation".

Can't recall that a little levity ever caused permanent cell damage, however lets get on with the project. Refer to CLEANING SECTIONS (19-1 & 19-2) of the book and use the same method. The importance of making certain that no abrasives or residue from Final Sanding-2 are carried over to the next cycle cannot be stressed enough!

Contamination from this cycle can easily cause damage in the Pre-Polish and Polish cycles, resulting in poor polish results, sometimes none at all, especially on materials below 7 in hardness\toughness on these scales. (See sections 18-1 &18-2)

# INTERMEDIATE TREATMENT

## GEMSTONES

Inspect your load...look for materials that have sub-surface fractures, exposed by reduction in size or shape in the Grind and sand cycles. ***Now is the time*** to remove this material!

---

*"it will be almost impossible to remove compounds from internal fractures without re-tumbling"*

---

As Pre-Polish and Final Polish are completed, either of these cycles in which fine compounds are used, especially in wet processes, the compounds may be forced into the fractures. This will create very "ugly" appearing baroques, pre-forms. or shapes...and it will be almost impossible to remove compounds from internal fractures without re-tumbling.

If your load contains materials below 7 in either hardness or toughness, a pre-polish cycle should be considered, before processing in the final polish. Remember...if your tumbler load is mixed as noted, tumble process for the softest and/or poorest toughness. Process the entire load for the most difficult to finish material in the load.

The harder materials will normally always finish with the least amount of difficulty, the softer materials will require the extra treatment. For example, pre-polish, longer running times in the final stages, finer abrasives, special cleaning and sorting between cycles...plus load cushioning.

As the final steps of the tumbling cycle are completed...there will be very little if any, further reduction of the physical size or shape of materials in the load.

# GEMSTONES-GEM MATERIALS

## ROTARY

| GRIT SIZE | MATERIALS | HARDNESS |
|---|---|---|
| 600 [Aluminum Oxide] | Amethyst, Apache Tears, Garnet, Glass, Jade, Jasper, Lapis, Opal, Peridot, Turquoise | 5--7 |
| Aluminum Oxide | 1 1/2--2 Tablespoons or 1 1/2–2 Ounces per lb. of load | |
| E-266, SF-1 | Feldspar, Zircon | 6--7 |
| Aluminum Oxide | 1 1/2--2 Tablespoons-- or 1 1/2--2 oz. per pound of load | |
| Load Capacity | 60-70% of Barrel Capacity Load levels are critical on softer materials..use pre-tumbled media or materials of same hardness | |
| Water Level | Just below final layer of materials. | |
| Run Time | 4-6 days (average processing time) | |
| Venting | Check after 1-2 days of run time to relieve pressure slowly if required. | |

Thickening, cushioning or suspending additives are usually required for loads containing materials below 7 in hardness or toughness. For processing materials below 5 in hardness such as Amber, Coral, Ivory, Nuts, Pearl, Seeds & Shell, see special section covering treatment for soft materials, Plastics, and Organics.

# GEMSTONES-GEM MATERIALS

## VIBRATORY

| GRIT SIZE | MATERIALS | HARDNESS |
|---|---|---|
| Aluminum Oxide 600 or SF-1 | Amethyst, Apache Tears, Feldspar, Garnet, Glass, Hematite, Jade, Jasper (Picture Type), Lapis, Opal, Pearl, Peridot, Turquoise, Zircon, Zoisite | 5--7 |
| 600--1000 | Amber, Bone,, Coral, Ivory, Organics, Seeds, Shell | 2--4 |
| Aluminum Oxide | 1--1 1/2 Tablespoons or 1--1 1/2 Oz. (by volume) per lb. of load. | |
| Load Capacity | 70%--80% capacity of container. | |
| Water | Add water equal to abrasive amount. | |
| Cover | Use clear kitchen wrap (multiple layers), stretch over container opening to retain moisture...place factory cover over top. | |
| Slurry | Maintain batter like consistency during processing....add water as required. | |
| Run Times | 1-2 days depending on materials. | |
| Cushioning | Thickening or Suspending additives recommended for materials 4 & below in hardness or toughness. | |

# GEMSTONES-GEM MATERIALS

## ROTARY-VIBRATORY

Why pre-polish? Good question...and one of the most frequent ask during demonstrations and seminars on tumbling. You'll note that pre-finishing is usually recommended only when finishing loads of softer materials and/or loads that contain softer materials. Many ask.."how does the tumbler know what to do"? The machine doesn't...it will just keep on turning the barrel or rotating the load, that's why you must provide the ingredients that will gently prepare the surface for polishing... one of the final steps in tumble processing.

Pre-polishing or pre-finishing can be the most important cycle in the tumbling procedure owing to the intensive treatment softer materials (6 & below in hardness--or toughness) receive in Grinding and Sanding cycles. The depth of the scratch pattern (pitting) increases, as the hardness and toughness of materials decreases. The harder--the tougher, the material...the shallower the pattern, the softer the material...the deeper the pattern.

The key is to use abrasives that reduce the high parts of the pattern...without deepening the pattern. Aluminum oxide (often referred to as ***alundum***) seems to provide a good balance. The reason being that fused Alundum has a fracture pattern (the way it splits or breaks down) different from silicon carbide. The same grit or grain size can be used, and each will produce different results, on the same load. Alundum is a tan colored abrasive, compared to slilcon carbide, which is black or gray. Alundum also fractures and wears with a more rounding surface, while silicon carbide fractures or breaks down with a sharp pointed surface.

# GEMSTONES-GEM MATERIALS

## ROTARY-VIBRATORY

One sequence for materials of 5-6 in hardness or toughness is to follow the #600 Silicon Carbide cycle (in Sand-2) with a similar cycle in pre-finishing, only using Alundum 600 (Al-6) as the abrasive. Remember... the materials, properties (hardness-toughness-shape), type of processor, size or capacity of the container, shape of the container, all have a bearing on abrasive size or grit size, processing times, and whether cushioning or thickening agents are required. There is no one formula or set of directions that cover every variation. This is why it is important to keep accurate well-organized records for each type of load. For some materials, temperature, liquid PH (acidity or alkaline) can be a variable factor. In many areas as water contamination increases it is suggested that distilled or bottled water be used, in place of tap water.

The important point to remember is that the surface of softer or materials which are not quite as tough as quartz or agate, may just polish faster, easier, and with a much higher gloss if you use a pre-finish cycle. Each tumbling cycle uses finer and finer (smaller & smaller) abrasive grain sizes...so if polish cycles are attempted, before the surface pattern is reduced to the proper depth....polishes will appear "hazy" or dull. The object is to provide a surface with a scratch pattern which not only gets progressively shallower, but one where the flat areas of the surface begin to overpower the pitted areas. Once this occurs, the polish and burnishing cycles can produce a finish that is highly reflective, and one that seems to have that "water wet" appearance. Skipped cycles, reduced tumbling times, eliminating pre-polish, thickening or cushioning, where applicable, may result in poor finishes!

# GEMSTONES-GEM MATERIALS

## ROTARY

In the final stages of tumble finishing, using rotary type machines, each processing cycle has averaged about one week. Counting the pre-finish cycle, about 5 weeks running time, several steps of preparation, loading and un-loading, cleaning between cycles, and experiences, which will prove invaluable during subsequent tumble processing...have brought your load of materials ready for the final steps in the tumbling process...Polishing and Burnishing.

---

*"enhanced beauty and pattern...each piece and individual and unique gem"*

---

Polishing is the final surface flow of materials to a high gloss or reflective finish. Translucent or clear materials gain depth and clarity, opaque materials reflection and color, both with enhanced beauty and pattern...each piece an individual and unique gem.

During the next few days all the time and effort of the past few weeks will come to an end. Selection of polishing compounds, cushioning agents, media, load levels and running times, will transform the materials into finished baroques, gemstones, and shapes ready for making finished jewelry, decorative items and specimens. This is the cycle that produces the rewards, that makes it all the more important that polishing operations are carefully and correctly executed. There is little room for error!

# GEMSTONES-GEM MATERIALS

## ROTARY

It's decision time. Now comes the finial stages of finishing. Thickening.....Media.....Polishing Agents.....Liquids...... Additives Running Times. To add or not to add. The materials in the load, by hardness, toughness, shape, and size, will all play a part in making the selections.

Remember...Loads levels must be sufficient to promote even tumbling action, too high and little action will occur...too low and loads will slide instead of tumble. As the slurry mixes in the final polishing steps are thinner than previous stages, total contents of the barrel are even more critical. Movement within the load has less resistance which, means fracturing or spalling can occur more rapidly.

Thickening of the liquid or adding cushioning may be necessary, especially if materials 6 & below are in the load. This can be accomplished by thickening liquids using common household items such as corn starch, gelatin, sugar, or syrup. Messy but effective. Additives such as wood pegs, walnut shell, corn cob, pecan shell, apricot shell, rice hulls, oat meal....again, some can result in mixes which are not too palatable, in appearance or smell...but have been used effectively in barrel tumblers to bring softer materials to a lustrous finish. Additives are sometimes referred to as "carriers" and confused with media.

Media are usually used throughout all the tumbling cycle...start to finish...while thickeners and additives are normally employed only in the final stages of pre-finishing and final polishing. Loads below 6 in hardness-toughness or materials of this category are prime candidates for the use of thickening, cushioning, or additives.

# GEMSTONES-GEM MATERIALS

## VIBRATORY

Vibratory cycles have averaged about 2 days each, totaling about 1 week, for completion of the Grinding, Sanding and pre-finish stages. Compressing the steps into a very short period compared to rotary (barrel) finishing. Vibratory finishers reduce the total times required, but not the amount of effort expended in processing the materials. Each stage must still be completed correctly, with sufficient processing times to process materials thoroughly during every step of the tumbling cycles.

---

*... "water wet" appearance..... sort of a self proclaimed standard, by which tumbled stones are judged.*

---

Assuming the operations completed have prepared materials correctly, the final steps in finishing the load will bring the materials to a final luster, in the polishing and burnishing cycles which are to follow. Depth, clarity, and color, will all be materially improved during the final processes.

Polishing and burnishing stages during the 2-4 days will transform the materials into finished gemstones, baroques, cabochons, shapes and gems, suitable for jewelry fabrication, specimens, and decorative items. The end results of subsequent steps will be to "flow" the surface of the materials, producing a highly reflective finish. The pursuit of a "water-wet" appearance, as if the materials have just been dipped in water, sort of a self proclaimed standard, by which tumble polish stones are judged.

# GEMS-GEM MATERIALS

## VIBRATORY

Material types, sizes, shapes, hardness, toughness and the proportion of these properties will affect the choices in selecting polishing compounds, medias, carriers, liquids, additives, tumbling intensities and running times. Sound complicated...sort of like modern day wood finishing. In the "old days" you could Oil Finish, Shellac, or Varnish...today...choices of woods finishes & colors are almost unlimited. Minwax® wood finishes, Polyurethanes, Polyshades®, Polycrylic®, Watco®, Helmsman®, Polyshades®, Woodsheen®, Antique Oil, Tung Oil, Danish Oil, Blend-Fil®, and this is after finishing with #80, #120, #220 and 000 Steel wool. (Like steps in gem finishing)

In the "old days" agates & jaspers..."into the barrel". Now its Amber, Agates, Amethyst, Apache Tears, Aquamarine, Aventurine, Apatite, Augite, Basalt, Beryl, Black Diamond, Black Opal, Bloodstone, Californite, Carnelian, Chalcedony, Chert, Chrysoberyl, Chrysocolla, Chrysoprase, Citrene, Coal (jet), Corundum, Dolomite, Emerald, Epidote, Feldspar, Greenstone, Helitrope, Hematite, Idocrase, Jasper, Jade, Jadeite, Kukui Nuts, Kunzite, Labradorite, Lazurite, Malachite, Melanite, Meteorites, Milky Quartz, Moonstone, Morganite, Moss Agate, Nephrite, Obsidian, Onyx, Opal, Peridot, Psilomelane, Quartz, Rhodonite, Sapphire, Shattuckite, Shells, Smoky Quartz, Sodalite, Sunstone, Tigereye, Topaz, Tourmaline, Turquoise, Uvarovite, Variscite, Wood Opal, Zeolites, Zebra Agate, Zircon, Zoisite, and all the variations with local names. But, like the varieties of woods and finishes...the end results are worth it. Vibratory finishers, Gem and Jewelry processors have opened up a whole new world of finishing.

# GEMSTONES-GEM MATERIALS

## VIBRATORY

Polishing compounds (like gem materials) are available in more and more varieties. The basics...Cerium Oxide, Chrome Oxide, Iron Oxide, Tin Oxide, & Tripoli (natural compounds are still widely used in barrel finishing. However these find only limited usage in vibratory finishing. Vibratory finishers produce both rotation and vibration, the hopper or container, instead of rotating, remains almost motionless. Abrasive grains are not only rotated, they are crushed at 1500 to 3500 times each minute... by vibration...in addition to the rotating action.

---

*"abrasive particle must break down slowly"*

---

Vibratory finishers produce both rotation and vibration. Hoppers or containers remaining almost motionless, abrasive grains are crushed gently but effectively at 1500 to 3600 times each minute by vibration...in addition to rotary action. These processes "break down" abrasives grains (polishes) very, very rapidly. Softer natural compounds tend to become too small...too quickly. All too often. the end results are not satisfactory. The answer lies in man-made compounds, and powders produced by fusion or sintering processes, combining oxygen and synthetic materials, during sintering--fusing procedures.

These processes produce compounds that are extremely well graded for particle size, and have the friable properties that the abrasives need to initially withstand vibratory action, slowly "breaking down" into smaller particles. Each tumbling stage has a progression during the cycle that must be completed, to produce the desired results.

# GEMSTONES-GEM MATERIALS

## VIBRATORY

Polishing compounds manufactured from Aluminum Oxides, include materials with identifying names and acronyms such as: Levigated Alumina, E-266, E-330, E111H, Cerium Substitute, SF-1, SF-2 just to mention a few. The point to remember is that Aluminum Oxide polishing compounds tend to withstand compressed vibratory tumbling times and action. By whatever name, or trade name, the polish is sold, ***synthetic and man made compounds are recommended for use in vibratory finishers.***

---

*"Aluminum Oxides...perform best on materials 5-8 in hardness/toughness"*

---

Synthetic sapphire compounds such as Linde-A® and Linde-B®, are also commonly used in vibratory polishing. In addition there are many new compounds appearing, almost on a monthly basis, most of which are manufactured from aluminum oxides. Aluminum oxides used in wet polishing processes will perform best on materials from 5-8 in hardness/toughness. What brand works best in each make of vibratory tumbler are usually set forth in the manufacturers operating instructions.

When in doubt, check with dealers, distributors, manufacturers, for recommendations on processing unusual materials, (below 5 or above 8 in hardness/toughness) shapes, or sizes. Your best references may be gained by consulting someone having experience with unusual materials, and by keeping accurate and detailed records of tumbling operations, procedures, compounds and additives.

# GEMSTONES-GEM MATERIALS

## VIBRATORY-DRY POLISHING

Dry Polishing...an alternate method of polishing a wide range of both soft to hard gem materials also including organics and metals. Wet'nDry *DEEP LUSTER*™ polishes use aqeuos (water based) liquids in a broad range of abrasive particulate sizes suspended into a creamy liquid base. This allows the impregration and charging of organic medias such as; Ground de-hydrated sized granules of corn cob, shell particulates such as walnut, pre-shaped hardwood and other medias The end results are Wet'nDry charged polishing compounds with a unique dual action.....and the best of all this is a cost effective method where the concentration of polishes are variable in accordance with your tumbling requirments.

***Wet'nDry polishing compounds produce that shine referred to as DEEP LUSTER™ in a new U-make it kit.***

Wet'nDry *DEEP LUSTER*™ compounds are bio-degradable. They contain ***no petroleum based chemicals.*** You mix and charge the dry medias right in in your own vibratory tumbler, as required. They are reusable, you can even clean and re-use your medias if so desired. The have a *dual action*. Once completely mixed and dried, only a slight amount of water is required to activate and operate in a semi-dry state. Operation is in a very slightly damp state until a completely dry environment is evident which causes the compound *to react with an even finer finish*. These pre-polishes and polishes are especially effective in finishing a wide range of soft or difficult to polish organic materials such as; Corals, nuts, shells and bone. Most all gem materials respond well including amber, lapis, mobe pearl, moonstone, opal, as well as harder gems.

# GEMSTONES-GEM MATERIALS

## VIBRATORY-DRY POLISHING

### *Wet'n Dry*™ ADVANTAGES

± Apache Tears & softer softer materials polish easily.
± Carriers and Medias eliminated.
± Additives, thickeners, cushioning eliminated.
± Chipping and Spalling...virtually eliminated.
± Burnishing process eliminated.
± Mixed materials easily processed.
± Liquid and slurry maintenance eliminated.
± Compounds migrating into fracture areas are reduced.
± Operating noise dramatically reduced.
± Wide selection of media base materials.
± Polishes most organics and Gem Materials.
± Re-usable and can be re-juvinated.
± Cleanup requires only a rinse in water.
± Messy compounds and slurries eliminated.
± Instant, clean examination.
± Pre-Polish Mesh sizes **800, 3000, 9000,**
± Polish Mesh Sizes **18,000 and 60,000**

The advantages of meeting individual requirements, compound concentrations, mixing and charging as usage requires, plus the *Dual action* of one mesh size reacting as if the load were processed in ***two*** mesh sizes of abrasives, while using just a single mesh size, far offset any disadvantages. The ability to always have freshly mixed dry polishes available, plus compounds that are storable and reusable with easy water cleanup for processed loads and the tumbling containers.

Wet,nDry ***DEEP LUSTER***™ compounds are available from 20 micron to 0.2 micron particle sizes. Dry polishing processes have been tried tested and proven world wide since 1965.

# GEMSTONES-GEM MATERIALS

## ROTARY

| TYPE-MESH | MATERIALS | HARDNESS |
|---|---|---|
| Cerium Substitute, Alum. Oxide .8 to 1 micron [Tin , Cerium, Iron Oxides] | Agate, Amethyst, Beryl, Chalcedony, Jasper, Quartz, Topaz, Tourmaline.<br><br>1-2 oz per. lb of load | 6--8 |
| Cerium Substitute, Chrome Oxide, Cerium Oxide | Apache Tears, Feldspar, Garnet, Glass, Hematite, Jade, Lapis, Opal, Peridot, Turquoise, Zircon. | 5--7 |
| Load Capacity | 60-70% of Barrel Capacity. Load Levels are critical, especially,with, softer,materials. Use 30-50% Media, or pre-tumbled materials of same hardness. | |
| Water Level | Just below final layer of materials | |
| Run Time | 4-6 days (Average processing time) | |
| Venting | Check each 1-2 days--relieve slowly. | |

Thickening, cushioning or suspending additives are usually required for loads containing materials below 7 in hardness or toughness. Materials below 5 in hardness such as: Amber, Coral, Ivory, Pearl, Seeds, Shell. (Special Section)

# GEMSTONES-GEM MATERIALS

## VIBRATORY

| TYPE-MESH | MATERIALS | HARDNESS |
|---|---|---|
| Wet-Method using: Cerium Substitute, Aluminum Oxide .8--1 Micron<br><br>*Materials must be coated..use only enough compound to make coating!* | Agate, Amethyst, Apache Tears, Beryl, Chalcedony, Chert, Citrene, Dolomite, Glass, Hematitie, Jasper, Obsidian, Peridot, Quartz, Rhodonite, Sodalite, Tigereye, Topaz, Tourmaline, Tourquoise, Zebra Agate. [See special section for materials below 5 in hardness]<br><br>1/4--1 1/2 oz per lb. of load [Check manufacturers instructions for data] | 5--8 |
| Load Capacity | 70%-80% capacity of container. | |
| Liquids<br><br>*If surfaceof the material is visible &/or foamy slurry mix is too thin.* | Initially start with liquid equal to the amount of abrasives by volume. Add additional liquid (if required) to create a slurry of a thick batter like consistency. Excessive liquid will make slurry too thin & foamy. | |
| Run Times | 2-4 days depending on manufacturers, instructions, and materials. | |

Thickeners, cushioning or suspending additives are usually required for materials 4 & below in hardness-toughness. Cube sugar 1---- 4 cubes per lb of load works well. Dishwasher detergent gel, added [Approx 1/4 oz. per 4 lbs of load] enhances tumbling action. Use Baking Soda or Sal Soda to control foaming, if this occurs. Kitchen wrap cover.

# GEMSTONES-GEM MATERIALS

## ROTARY RE-CAP

The old axiom, "stones are made in the sanding", is one to remember. However sans correct polishing...all efforts are without significance. Lets re-cap:

### 10 POINTS--RULE OF THUMB--10 POINTS

#1. Polishing--the final step in surface flow of materials.

#2. Decisions on polish compounds, medias,, thickeners, liquids, additives & running times must be made.

#3. Shape, size, hardness/toughness of materials affects selections. The higher the ratio of shapes, pre-forms, flat surfaced, and larger items...the higher the ratio of media to load...[up to 50%--60% media]. Mixing materials below 5 in hardness/toughness is not recommended, without previous experience.

#4. Load levels can be critical, especially with softer materials, shapes, pre-forms, & flats.

#5. Thickeners/additives...consider with loads containing materials below 6 in hardness, shapes, & pre-forms.

#6. Check load for excessive noise. Additional media, thickeners, additives, & cushioning may be needed. Commercial products thicken, improve slide and suspend abrasives for improved tumbling action.

#7. Foaming may require addition of baking soda or sal soda, or commercial de-foamers.

#8 Venting may be require periodically to relieve pressure.

#9. Dishwasher soap [or gel] tends to promote tumbling action. Also produces lessd foaming than dish soap.

#10. Cube sugar sometimes used as thickening agent, if no commercial additives are available.

# GEMSTONES-GEM MATERIALS

## VIBRATORY RE-CAP WET METHOD

Vibratory tumblers are more "forgiving". Nevertheless because tumbling times are compressed, the correct procedures become even more important.

## 10 POINTS--RULE OF THUMB--10 POINTS

#1. Complete stage correctly, with sufficient time to remove texture from pre-polish...process thoroughly.

#2 Material types, shapes, size, hardness/toughness, & ratio all affect the polishing selections.

#3. Aluminum oxide and synthetic sapphire polishing compounds recommended as first choice.

#4. Media to load ratios highest when loads contain materials below 6 in hardness/toughness, shapes, pre-forms and flats. Processing materials below 5 in or above 8 hardness...with previous experience.

#5. Thickeners/additives should be considered when materials are below 5 in hardness/toughness...with loads containing shapes, preforms, or flat surfaces.

#6 Load levels 70%--80% capacity of container for best tumbling action and results.

#7. Foaming may require addition of baking soda, sal soda, or commercial additives.

#8. Slurry consistency should be maintained in thick batter, or thin paste. Materials should be evenly coated, if surface is visible...slurry is too thin...or insufficient compound.

#9. Dishwasher [Gel]...small amount can improve tumbling action. Also produces less foaming.

#10. Cube sugar (or propolyene glycol) can be added as thickener in lieu of commercial additives.

## GEMSTONES-GEM MATERIALS

### VIBRATORY DRY POLISHING

Dry polishing is the clean, easy way to superbly polished, organics, bone, fossil ivory and gemstones. In two to three short cycles, on the average, most tumbled loads are completely polished.

***SUPER PRO DEEP LUSTER***™ are highly concentrated abrasive compounds formulated into a creamy liquid. Mixed with a dry media such as; Milled Dehydrated *Corn Cob*, *Walnut Shell* or *Wood cubes*, the result is ready to use dry pre-polish or final polish. You choose the dry media of your choice, select the compound mesh size; 800,3000,9000,18,000 or 60,000 and in just a few minutes mix and charge the media for pre-polishing or final polishing, using your own tumbling container hopper or tub. The compounds may also be pre-mixed and stored in containers for future use.

To use Wet,nDry your container should contain; 1/3 Wet'nDry premixed, 2/3 tumbling load, with the hopper, tub or container approximately 70-75% full.
Start the tumbler, *"wet"* the load with just enough water to dampen the mixture and activate the abrasives. ***Apply the water very sparingly.*** Too much water will cause the abrasive to lose its effectiveness.

Using a plastic wrap such as Glad® Press'nSeal around the edges of your container, create a moisture retaining seal to avoid evaporation for the first few hours (36-48). Remove the seal and allow the mix to complely dry (12-24 hours). Repeat for each step until the load is polished. Times vary with gem materials.

***DEEP LUSTER***™ contains *no petroleum based oils or lubricants*. Water may be usedr as an extender fluid. Rinse and clean the finished load and tumbling container using water. Compounds are bio-degradable.

# GEMSTONES-GEM MATERIALS

## VIBRATORY DRY POLISHING

### DEEP LUSTER™ Wet'nDry Polish Cycle

| MESH | | SUGGESTED USAGE |
|---|---|---|
| #800<br>#3000<br>#9000<br>#60,000 | Step #1<br>Step #2<br>Step #3<br>Step #4 | **4 Step Processing Cycle**<br>**Materials--2--4 in hardness**<br>**EXAMPLES**<br>**[Amber, Coral, Pearl]** |
| #800<br>#3000<br>#18,000 | Step #1<br>Step #2<br>Step #3 | **3 Step Processing Cycle**<br>**Materials--4--5 in hardness**<br>**EXAMPLES**<br>**[Apache Tears, Glass, Shell]** |
| #3000<br>#18,000<br>#60,000 | Step #1<br>Step #2<br>Step #3 | **3 Step Processing Cycle**<br>**Materials--5--6 hardness**<br>**EXAMPLES**<br>**[ Jade, Opal, Turquoise ]** |
| #3000<br>#18,000<br>#60,000 | Step #1<br>Step #2<br>Step #3 | **2 Step Processing Cycle**<br>**Materials--7--8 hardness**<br>**EXAMPLES**<br>**[Agates, Jasper, Topaz ]** |

**DEEP LUSTER™** Wet'nDry
Mesh Sizes Available

| | |
|---|---|
| **#800** | **Pre-Polish** |
| **#3000** | **Pre-Polish** |
| **#9000** | **Pre-Polish** |
| **#18,000** | **Polish-Final** |
| **#60,000** | **Polish-Final** |

**Ultrafine Deep Lusterxxx**
**Bio-degradable**
**Creamy Liquids**

Use 1½-2 oz. **Wet'nDry** Compound [by weight] for each pound of load. Approx. 25%-30% of the load [ **by** volume] should be Wet'nDry™. Softer materials require more steps, harder materials less.

*Too much compound may create excess residue on stones & container.*

# GEMSTONES-GEM MATERIALS

## ROTARY

Burnishing can be implemented to enhance luster on some gem materials. Surface may appear slightly hazy, as if a "film" is present, or wetting with water makes the surface appear clearer and more reflective.

*" One lump or two.....burnishing with soap, water and sugar...can produce fantastic results"*

This condition may occur as a result of "short changing" one of the previous cycles. Or, contamination in the pre-polish or polishing cycles. *Improper load levels in one or more previous steps, can also result in stone to stone contact, causing very minute sub-surface fractures.* These fractures are in turn exposed *by surface flow* in the pre-polish and polish stages, producing a surface that appears to have dull "film".

Another culprit, which may go unnoticed, is materials in the load that have split or fractured during final sanding, pre-polish, or polishing cycles. Small particles of this material can act as an abrasive, dispersing throughout the load, resulting in very shallow pits. A surprisingly small amount of contamination can cause considerablef damage.

Burnishing, using a mixture of powdered soap, water, and sugar, is one method. One lump or two per pound of load can produce fantastic results. Consistency of the burnishing slurry seems to be a key factor for best results. Burnishing times are usually short, require periodic monitoring, processing can result in surface enhancement.

# GEMSTONES-GEM MATERIALS

## ROTARY

### POWDERED SOAP METHOD

Clean load thoroughly and inspect for any materials which might be porous, split. or fractured and remove from the load. Drain off all water that runs freely, and load barrel.Add or remove pre-tumbled gem material or media for proper load level.

**Soap** Start by addition of 1/2 ounce per lb of load Ivory Snow®,Ivory® Flakes or commercial burnishing Soap. (Add more if required) Royal Green liquid burnish soap can be used in a thinner consistency, with ceramic media, with ***excellent*** results. (Available from Custom Technology Ltd.)

**Water** Add just enough water total, to make a slurry like consistency. Just thin enough for the load to move and freely tumble.

**Mixture** If the mixture is too thin, stone to stone contact will occur, causing more harm than good. Commercial soaps are recommended

**Additives** Cube sugar (for smaller barrels 1--6 lb capacity) is easy to measure. Use granulated for larger containers.

**Duration** Anywhere from minutes to hours. Check load frequently. Don't over burnish.

# GEMSTONES-GEM MATERIALS

## VIBRATORY

Burnishing with vibratory tumblers can enhance the appearance of some types of material that emerge from the polish with a slightly hazy or subdued reflective surface. A condition that may result from any one of several causes: load level, intensity, consistency, contamination. Improper polishing compounds, "short changing" one of the previous cycles. Small particles of material breaking away due to fracturing or spalling, can likewise produce similar conditions. Whatever the cause, the effect is usually the same: A surface appearance that needs improvement.

The method can be as time tested as soap, sugar and water, or utilize one of the more recent commercial burnishing compounds and additives. Whichever or whatever is used, the objective mixture or compound has four pre-requisites: It should:

#1. develop a Thin slurry consistency.
#2. develop a Slippery, Slick, environment.
#3. **envelop** the load completely.
#4. prevent any bare stone to stone contact.

Loads that have been incorrectly processed, or surfaces which have a poor polish, are not candidates for burnishing treatment. These should be re-processed in whatever cycles are necessary to bring to a polish...burnishing will have little effect. Burnishing is a supplement, not a substitute for polishing. The goal is enhancement, improvement, refinement...perfection of what has already been completed over many hours of processing.

# GEMSTONES-GEM MATERIALS

## VIBRATORY

## POWDERED SOAP METHOD

Vibratory finishing mixtures must be in a thin to thick slurry consistency. Wash load, sort out any suspect materials that could split, or carry contamination, **drain off all liquid that will freely run.** Load level should be 70%-80% total capacity of the container, well balanced between large/small materials and include media used in the polish cycle. Do not add any media or material that has not been through the polish...this could result in surface damage, and require re-finishing of the entire load.

---

*"One distinct advantage with vibratory finishers...open containers...mixtures can be developed visually..."*

---

A sudsy mix could spell disaster, [we're not washing dishes] this is the last and final improvement...if the load warrants it. A super concentrated soap mix will form a burnishing slurry. One distinct advantage with vibratory finishers are open containers...mixtures can be developed visually to the correct consistency, without guess-work.

Once the burnishing consistency is developed, stretch kitchen wrap over the opening, and hold in place with a rubber band to preserve moisture. The rotation should be sufficient to keep the load turning. Use dial speed/intensity adjustment if available. A liquid burnish soap, (Royal Green) available from Custom Technology Ltd. does a fantastic job when used with ceramic media.

No appreciable material is being removed. Enhancing reflectivity and clarity, are the primary goals in burnishing.

## GEMSTONES-GEM MATERIALS

### VIBRATORY

Start with a load that has been washed, inspected, and drained of excess liquid, but wet. Load level....70%-80% capacity in a container which is free from contamination. [Do not use containers contaminated or embedded with silicon carbide in the walls or bottom] A sound policy is to reserve one container for polishing--burnishing only.

| | |
|---|---|
| Soap | **Start with 1/2 ounce per pound of load [More can be added later]. Add with load in rotation.** |
| Water | Add 1/4 oz per pound of load to develop thick layer of foam. [Any additional water should be very, very sparingly] |
| Mixture | Stone to stone contact is to be avoided. Add additional soap if this condition is evident. If there is excess liquid initially, sudsing may occur. Surfactants are required in the form of foam. A thin mix with high suds is undesirable. |
| Additives | Cube sugar...1--2 lumps per pound of load can thicken...impart a "gripping" effect. |
| Duration | Processing can take from minutes [under 1 hr] to several hours. Depending on materials, shape, size. Over burnishing can result in damage, check hourly. Enhancement goals are to: reduce haze, improve reflectivity, "depth" and color, not to re-polish the materials. |

# GEMSTONES-GEM MATERIALS

## RE-CAP

Lets re-cap the burnishing process and the objectives.

**First**......Measurments for soap, water and additives, are not "written in stone".

The reverse however is quite true, as many times stones do dictate the measurements. Softer materials require thicker mixes than hard, and burnishing with shorter cycles. Use only a Commercial Burnishing Soap mix on materials below 5 in hardness.

**Second**.....There's more than one way to burnish! The old adage, "there's more than one way to skin a cat" certainly applies to burnishing. Softer materials may in some instances be burnished in thinner consistency commercial mix using stainless steel shot, balls, ball-cones, diagonals, and pins along with commercial burnishing soap. Check manufacturers instructions, and suppliers for additional compounds, information, and listings.

**Third**.......The burnishing objectives are to improve surfaces. Enhancement, to reduce surface "haze". Improve reflectivity, clarity, "depth", and color.

**Fourth**.....Burnishing does not remove any appreciable amount of material. The process is not a substitute for polishing. Intense speed and rotation are not pre-requisites.

**Fifth**.....It is always the best policy to test the procedure with average materials before using on expensive loads.

# GEMSTONES-GEM MATERIALS

## ROTARY

The basics for Shapes and Pre-forms are about the same, some softer materials may require different abrasives.

| | |
|---|---|
| Grinding | Pre-shape, pre-form and finish with 180-220 or finer diamond wheel; |
| Media | 40% to 60 % ratio of media to load. Intricate shapes require higher ratios. Varied sizes, & smaller media carries the abrasive into contours, separates the work pieces, evens treatment into all areas. |
| Load | Barrel machines 65%-75% capacity. |

| | ROTARY- SANDING [Step #1] | |
|---|---|---|
| ABRASIVE SIZE #600 | MATERIALS<br>Agate, Amethyst, Apache Tears, Garnet, Jasper | HARDNESS<br>5--7½ |
| Silicon Carbide | 1½--2 TBS. per lb. of load [1½-2 oz.] | |
| Water Level | Just below final layer of materials. | |
| Run Time | 4-6 days...[Check load every 2 days. Cycle completed, when all areas show equal treatment and uniform surface. Longer cycle required on flat surfaces. | |
| Venting | May be required after 4 days. | |

Thickening agent may be required when load contains materials below 6 in hardness/toughness.

# GEMSTONES-GEM MATERIALS

## ROTARY

| | PRE-FINISH [Step #2] | |
|---|---|---|
| Abrasive Size | Materials | Hardness |
| 1000 Silicon Carbide | Amethyst,PictureJasper, Garnet. | 6-7½ |
| AL-6 [ 600 ] Aluminum Oxide | Apache Tears, Glass, Opal, Turquoise, Zircon | 5-6 |
| Water Level | Just below final layer of materials | |
| Run Time | 3-4 Days [ Increase for flat surfaces] | |
| Venting | Check after 3 days | |

| | POLISHING [Step #3] | |
|---|---|---|
| COMPOUND | MATERIALS | HARDNESS |
| Cerium Substitute Tin Oxide | Agate, Chalcedony, Jasper, Quartz, Topaz, Tourmaline | 6½--8 |
| Cerium Oxide or Chrome Oxide | Apache Tears, Feldspar, Glass, Hematitie, Opal, Peridot | 5--6½ |
| Load | 60%-70% Barrel Capacity | |
| Water level | Just below final layer of materials. | |
| Run Time | 5-7 Days [Check progress every 2 days] | |
| Venting | Check after 2nd day | |

Thicken or cushion loads which contain materials below 6 in hardness/toughness. Flat areas require increased times.

# GEM MATERIALS--SHAPES PRE-FROMS

## VIBRATORY

Variable controls...electronic or mechanical...are a distinct advantage in processing shapes and pre-forms. Tumbling slurries are thinner, with less ability to cushion loads. Materials are susceptible to damage, especially on points of contact farthest from the center of the object. Mechanical advantage is greatest at this point and the edges farthest from the center, usually receive greater treatment.

This is especially true on preforms such as cabochons. The polish will occur first on the outer edges and work towards the center, especially on flat areas. Flat areas require longer processing and need to be separated to prevent "lapping" together and receiving little, if any, treatment. Odd shapes need uniform contact on all areas for equal treatment, and to prevent distortion of shapes.

The amount of media [40% to 60% by volume] in relation to the load is not uncommon, depending on shapes, sizes, and materials. (adjust for particular type load if available)

One of the most common mistakes in tumbling, seems to occur at this point. Processing of pre-forms and/or shapes, without media, is certain disaster. Especially in the stages that must be performed in "wet" cycles, with loose abrasives and water. This includes, sanding, pre-polish, polish and burnish steps. Pre-forms and shapes require special consideration in the selection of abrasives, compounds, media, tumbling intensity, and cushioning additives. Combination of various types and sizes of media can be used. One common technique is to use varied sizes & shapes of media, one to move the other.

# GEM MATERIALS--SHAPES PRE-FORMS

## VIBRATORY

Shapes and pre-forms retain closer tolerances to their original form with vibratory finishing. Gentle rotation coupled with vibrations of 1500 to 3600 per minute, effectively process materials. Rotation keeps the load evenly coated with abrasives and/or compounds, while vibration gently, but quickly finishes most materials to a brilliant polish. The speed of circular rotation is not critical to achieve results, especially when processing larger objects with intricate shapes. For best results, all contoured areas [especially concave] should be uniformly pre-formed. Media size and shape should be such that all areas of the work piece and media are in contact. Open areas with little or no contact will not receive effective and uniform treatment.

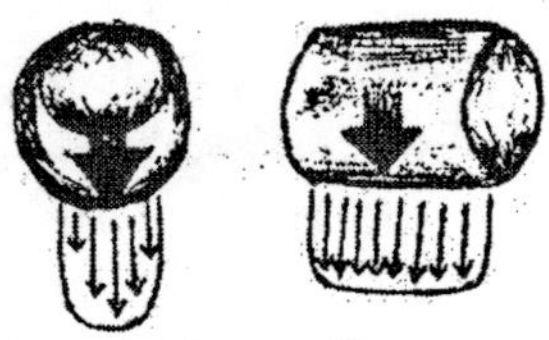

Cylinders have 50% greater effective area

Processing after wet final sanding, and pre-polishing, [if used], Wet'nDry™ compounds provide an extra measure of control. Not only is the process clean, but compounds also act as the carrier, media, and polishing agent, combined Into one. ***DEEP LUSTER***™ compounds require just a small amount of liquid to activate abrasives...compounds are then allowed to "dry" during operation. Deep Luster compounds are very useful on special materials that might discolor. (such as softer composites, delicate or intricate shapes, some types of turquoise, & some opals) Both types allow instant inspection of progress, without messy cleanup routines.

# GEM MATERIALS--SHAPES PRE-FORMS

## VIBRATORY

| | |
|---|---|
| Pre-Shape | Grind or pre-form, finish with 220 grit wheels or finer. Deep grooves or marks must be well blended or removed. |
| Media | Media must get into grooves & contours....use combination of shapes and sizes that best conform to shapes being finished. Contact is the key word in finishing shapes & preforms. |
| Load | Vibratory loads [including media] 70%-80% capacity of container. |

| ABRASIVE- SIZE | VIBRATORY- WET SANDING–1 [HARD] |
|---|---|
| #600 | MATERIALS — HARDNESS<br>Agates, Amethyst, Beryl, Citrene,Garnet,Peridot, Topaz, Tourmaline. — 6--8 |
| Silicon Carbide | 1½--2 TBS. [1-2 ounces] per lb. of load. |
| Water-liquids | Drain all that runs freely, add abrasive slowly. Check slurry, maintain batter like consistency..add liquid if required. |
| Running time | 1-3 days [check load frequently] Equal uniform treatment is the goal. Flat areas may require longer cycles |
| Additives | Thickening, cushioning, or suspending agents should be used on intricate shapes or softer materials. |

# GEM MATERIALS--SHAPES PRE-FORMS

## VIBRATORY

| ABRASIVE- SIZE | WET SANDING–1 [SOFT] |
|---|---|
| 800 Grit | MATERIALS HARDNESS<br>Feldspar, Glass, Hematite,<br>Lapis, Opal, Shell, Turquoise 4--6 |
| Silicon Carbide | 1½-2 TBS. [1-2 ounces] per lb. of load. |
| Water-liquids | Drain all that runs freely, add abrasive slowly. Check slurry, maintain batter like consistency..add liquid if required. |
| Running time | 1-3 days [check load frequently] Equal uniform treatment is the goal. Flat areas may require longer cycles |
| Additives | Thickening, cushioning, or suspending agents should be used on intricate shapes or softer materials. |

| Abrasive<br>SF-1, or 35-50 Micron Aluminum Oxide Compound 1-2 ounces per lb of load.<br><br>Additives to make slurry thick, and cushion load.Use commercial types | WET SANDING–2 [SOFT]<br>MATERIALS HARDNESS<br>Feldspar, Glass, Hematite,<br>Lapis, Opal, Shell, Turquoise 4--6<br><br>Drain load..add compound slowly to make thick batter like slurry. Maintain Coating...add liquid very sparingly if required. Load level of 70%--80% capacity. Runtimes of 1-3 days..check frequently. Flat areas require longer treatment. |
|---|---|

# GEM MATERIALS-*SHAPES PRE-FORMS*

## VIBRATORY

| COMPOUND SIZE | POLISHING–WET METHOD [HARD] |
|---|---|
| | MATERIALS HARDNESS |
| Aluminum oxide polish compound | Agates, Amethyst, Beryl, Citrene, Garnet, Peridot, Topaz, Tourmaline. 6--8 |
| Cerium Substitute or equivalent, in 1 micron or less | ½-1 Tbs [½-1 ounce] per lb. of load. Follow manufacturer's instructions. |
| Water-Liquids | Drain all that runs freely & quickly. Add compound slowly to rotating load. Add liquid very sparingly, only if required. Maintain batter like slurry. |
| Running Times | 2-4 days [check load frequently] Flat surfaces will require longer treatment. |
| Additives<br><br>[Note: Check machine intensity settings...adjust before operation. Use only enough intensity to evenly rotate load. Damage can occur quickly, if precautions are not observed!! | A small amount of **dish washer gel** can help break surface tension of water. Thickening--Cushioning agents should be considered on intricate shapes, large work pieces, or softer materials. Use commercial suspending agents when available.<br><br>Sal Soda© or baking soda can help deter sudsing, control excess foaming. This can dilute polish slurry, resulting in spalling very quickly. Slurry must "cling' to load, **not foaming or bubbling.** |

# GEM MATERIALS--*SHAPES PRE-FORMS*

## VIBRATORY-- DRY METHOD

After *completing* the *grinding* and *sanding* steps consider the new Wet'nDry™ method of polishing steps by using ***DEEP LUSTER***™ compounds for the pre-polishing and final polishing operations.

### POLISHING–DRY METHOD

| ***DEEP LUSTER***™ Mesh Sizes | MATERIALS | HARDNESS |
|---|---|---|
| 3000 & 18,000 Wet'nDry™ | Agate, Chalcedony, | 7--7½ |
| 3000-9000-18,000 Wet'nDry™ | Amethyst, Beryl, Citrene, Jasper, Garnet, Peridot, Tourmaline | 6½--8 |
| 3000-9000-60,000 Wet'nDry™ | Beryl, Chrysoberyl, Spinel, Topaz | 7½-8½ |
| 800-3000-60,000 Wet'nDry™ | Apache Tears, Feldspar Glass, Hematite, Shell, Turquoise [Natural] Zircon | 5--6 |
| 800-3000-18,000 Wet'nDry™ | Coral, Ivory, Lapis, | 3--5 |
| 800-3000-18,000 Wet'nDry™ | Opal, Turquoise [Stabilized] | |

***For a final Super Polish use Deep Luster Ultra xxx ™***

Processing cycle times will vary with the size and shapes of gems to be processed. These are guide lines only!

***Initially*** add sparingly enough *purified* water to create a ***damp*** state to the pre-mixed compound. 1/3 compound & 2/3 load to = 70% of container capacity. Seal tumbling container with Glad® press'n Seal. Run *2 days damp,* remove seal , run *1-2 days* until compound dries, for each mesh size. Adjust running times as required, for specific shapes, material sizes and mixed loads.

# GEM MATERIALS--FLAT SURFACES

## ROTARY

Flat surfaces [including slabs] require a measure of special attention in order to obtain highly reflective polished surfaces. ***Flats [slabs] need separation***, not only to keep the surfaces from "lapping" together, but to maintain the flow of fresh abrasive to the large surface areas. The ratio of media or smaller material used to fill in spaces may be in the range of 40% to 60 %. Small rounded material 1/4" to 3/8" makes a natural media for use in processing flat surfaces. Another is ceramic cylinders 1/8"--3/16" with angled ends. Properly balanced loads can support up to 30% flats without special media.

---

***"Properly balanced loads could support up to 25% flat surfaced material without adding special media."***

---

Most flats that have been diamond sawed receive about a 320 grit surface in the sawing process, especially if a "power feed" model has been used. Processing should not proceed in coarser than #400 unless saw marks are present. Starting the flats in Rotary Sanding-2 would be normal, Sanding-1 if saw marks are pronounced.

Surface diameter of individual flats should not exceed 25% of the barrel diameter, to keep tumbling action uniform within the load. Remember, the large surface area does not lend well to receiving full mechanical advantage, during the tumbling cycle. Flat surfaces, [slabs], and flat areas may require up to 4 times longer in the processing cycles. During sanding cycles it is difficult to detect insufficient treatment by reflectivity. Surface treatment starts on the edges & works towards center.

# GEM MATERIALS--FLAT SURFACES

## VIBRATORY

Vibratory processors alleviate some problems inherent in finishing flat surfaces [slabs], however the basics still have to be coped with. Surfaces have to be separated to prevent "lapping" together. Loads have to be balanced with 30%--50% smaller media for even processing.

Gem & jewelry processors equipped with variable speed controls [mechanical or electronic] provide a distinct advantage, especially with thin, delicate, intricate, large work pieces and softer materials. Control of intensity and rotation can in some instances eliminate the need for additives, cushioning, Vertical rotation, as opposed to horizontal, can be a plus in holding closer dimensions when processing intricate shapes and pre-forms.

Diamond-sawed flat processing can normally begin in the Sanding-2 stage, with # 500-600 silicon carbide. Loads with 30%-40% media by volume, and 60%-70% flats are average. *Use coarser silicon carbide only if grinding or saw marks are pronounced*, followed by 500-600 grit cycle.

Processing times for flat surfaces are three to four times longer than for rounded or spherical shapes. Allow for extended running times...if you cycle through the final finish with only the outer edges of flats receiving polish...this is a good indication of inadequate pre-treatment. Special dry compounds used in finishing flats are usually formulated using cubes, rectangles, or triangles as the abrasive carrier. These are especially useful on materials which undercut, or contain materials of more than one hardness such as; Jade, Lapis, Rhodonite.

# GEM MATERIALS FLAT SURFACES

## ROTARY

| ABRASIVE SIZE | MATERIALS HARDNESS |
|---|---|
| | Agate, Amethyst, Jasper, Quartz. 6--7½ |
| 320-400/F | ROTARY WET SANDING–1 [HARD] |
| 500\F--600/F | ROTARY WET SANDING–2 [HARD] |
| Silicon Carbide | 1½-2 TBS. per lb of load [1½-2 oz.] |
| Water Level | Just below final layer of materials. |
| Run time | 5--8 days......[Check load every 2 days.] Cycle is complete when even uniform surface is evident from edge to center. Processing cycles which are too short could result in partial polish around edges only...with reduced center polish. |
| Venting | May be required after 2--3 days. |

Thickening, Cushioning, Suspending agents recommended if loads contain materials below 6 in hardness/toughness, or large flats sections. Do not overload with large flat pieces as this could result in uneven tumbling action, with damage to edges of flat sections.

Sanding-1 cycle may be required only with loads containing materials below 6 in hardness/toughness...or if saw or pronounced grinding marks are present. A uniform even surface texture is the goal in these cycles. Finish tends to work from the edges to the center. Incomplete processing can result in poor polishing in the next cycles.

# GEM MATERIALS FLAT SURFACES

## ROTARY--VIBRATORY

| COMPOUND | MATERIALS HARDNESS |
|---|---|
| | Agate, Amethyst, Jasper, Quartz. 6--7½ |
| SF-1,—Aluminum Oxide 10 micron 600 Silicon Carb. | ROTARY WET PRE-FINISH [HARD] 1--2 Tbs. per lbs of load [1--2 0z] VIB. WET PRE-FINISH 1--2 Tbs. per lb of load [1-2 oz.] |
| Super Polish Aluminum Oxide ½–1 Micron or Cerium Subst. | *ROTARY WET POLISH* 1-2 Tbs. per lb of load VIBRATORY--1--1 1/2 TBS PER LB. |
| Water Rotary Water Vibratory | Just below final layer of materials. Maintain a thick batter like slurry. |
| Run time Rotary Run time Vibrator | 6-10 days (Check load every 2 days) 4-6 days (Check load each day) |

Thickening, Cushioning, Suspending agents recommended if loads contain materials below 6 in hardness/toughness, or large flats sections. Do not overload with large flat pieces as this could result in uneven tumbling action, with damage to edges of flat sections.

Pre-finishing may be required only for materials below 6 in hardness/toughness. Softer materials also require longer running times. Harder materials [7--8½] may polish first if being processed in a mixed load. Remember flat areas require longer running times. The larger the surface area of single pieces...the longer the cycle required to polish.

# METALS--GOLD & SILVER

There are some basic metal finishing procedures similar to those used on gem materials. Preparation of the surface for instance is of primary importance. As with any tumbling operation, satisfactory polishes are difficult, if not almost impossible, with improperly prepared surfaces. Pre-finishing is therefore usually required, and sometimes with more than one cycle and/or compounds.

---

*"final goals are highly polish reflective surfaces"*

---

As with any finishing operation there are of course always exceptions. Fabricated work from sheet type metals do not require nearly the amount of pre-finishing as would be used on "sand" or "cuttlebone" castings. Investment or lost wax silver or gold castings would require pre-finishing, the degree depending or the texture of the surface. Casting methods, temperatures, quality of the investment mix, and techniques can all have effects on the surface of the finished work piece.

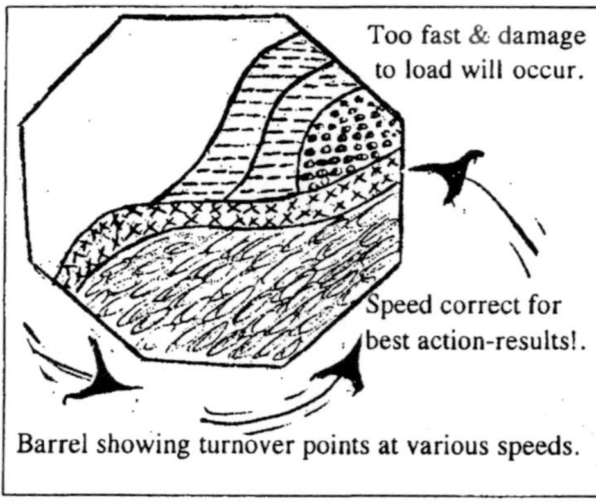

Barrel showing turnover points at various speeds.

Metal-finishing procedures should not be confused with Gem Materials, for which the object is usually large amounts of surface reduction, but there is a general similarity. The surface material removed during metal finishing cycles is small by comparison. Operations can range from wet processing with plastic, pre-charged organic, custom charged, and metal shapes.

# METALS--GOLD & SILVER

Polishing can be as varied as it is with pre-finishing, in fact, even more so. Finishing can be done with organic media such as pre-charged milled Corn Cob, ground shell [Apricot, Pecan, Walnut], Hardwood cubes, rectangles, and triangles, normally in dry-types of operation, or, wet-finishing using steel shot or ceramics in various shapes, balls, cones, ball-cones, ellipses, diagonals, pins, adding various types of soaps, compounds and water. In addition to these methods, various combinations of the above have been used effectively. Custom charging with special abrasives and compounds can produce excellent results.

---

*"...goals of metal finishing are smooth, uniform surfaces, highly reflective.....with 'eye appeal'."*

---

Metal finishing by any other name has acquired the antonomasia as one of the "black arts". It certainly may seem this way, especially if one is just beginning, processing something out of the ordinary, or trying to obtain information about special techniques or treatments. This section of the book is not intended to be a complete instructional concerning all phases of metal finishing, as this would require a book in itself.

The goals of metal finishing are smooth, even, uniform, surfaces, highly reflective and free of flaws, blemishes and imperfections, with "eye appeal". Retention of as much of the original intricate and decorative detail as possible is obviously desired. Tumbling operations seldom are used to remove appreciable amounts of metal, because the loss of detail would be prohibitive. Each cycle or operation should

# METALS--GOLD & SILVER CASTINGS

Clean and prepare surfaces removing and blending any sprues, imperfections, and fabrication marks. Use flexible shaft equipment, abrasive cord, files, sanding cloth, sanding sticks, rubber Cratex® wheels, or Scotchbrite® wheels.

Media size and shape should be selected so that lodging does not occur. Media should be small enough to pass through any opening in the work piece, or large enough so that it cannot stick in the opening of any detail. Free movement of the media must occur at all times during the processing cycle. Otherwise uneven treatment can occur, resulting in loss of detail on the work pieces. Regardless of whether media selected is organic, metallic, or ceramic, the same basic principles apply.

Vibratory units function well with organic, plastic, and ceramic types of pre-charged medias. Shapes and sizes of medias or carriers selected must be small enough to get into detail of the work pieces, but not become lodged during processing. You can spend more time removing lodged media than finishing each piece individually, especially with intricate detail or filigree work pieces.

# METALS--GOLD & SILVER CASTING

Pre-finishing and preparation using vibration can also be effective with custom charging or mixing of creams, gels, compounds and powders. Pre-charged media such as plastic should be pre-processed to "break in" the media, removing any sharp edges, prior to processing the metals. Finishing cycles tend to be shorter with vibratory units and considerably less liquid is employed in wet processing,

UV-18

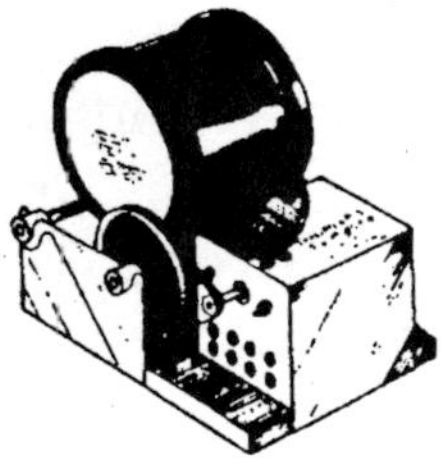

Model T

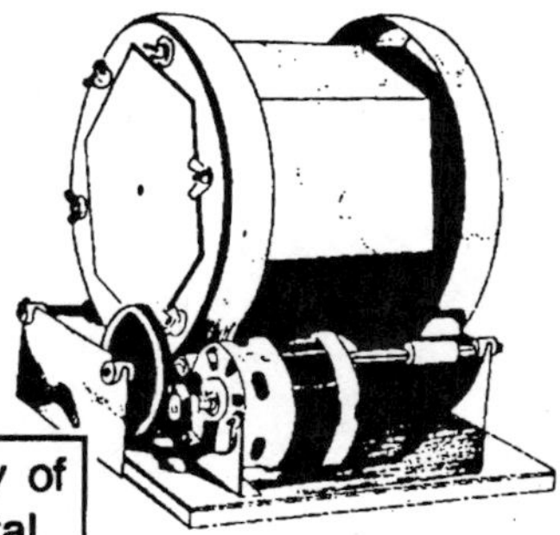

Model B

Photos courtesy of Tru-Sqare Metal Products

than in machines operating with "flow through" systems. Other types of systems using centrifugal, and combinations of rotary and centrifugal offer considerable reduction in processing times, plus enhanced retention of details.

Centrifugal "spin" finishers usually process materials in a **matter of minutes,** compared much longer times using barrel or vibratory equipment. The trade off is initial cost, however if production requirements warrant the expense, cost recovery can quickly offset the investment. Two distinct types of centrifugal units are classed as "Contact" and "No Contact". Both types basically function with the same principles of centrifugal force, only the methods of applying the force & containers differ, along with degree of control of the operations.

# METALS--GOLD & SILVER

*Contact units* spin the contents, which are loaded onto a flat disc/plate inside the container along with media and compounds. Centrifugal spin causes "G" forces to be applied to the work pieces in rotation. Very close tolerances must be maintained between the spinning plate and outer wall of the container.

*Non-Contact machines* function with magnetic drives situated below the work piece container. Only the contents (work pieces, medias, compounds and solutions)) are rotation, in the upper removable container. This allows a controlled type of environment, very important features for processing work loads requiring special treatment.

Magnetic barrel finishing machines process jewelry and small metal parts using thin stainless steel pins and balls. Burnishing to brilliant finishes with very little loss of detail.

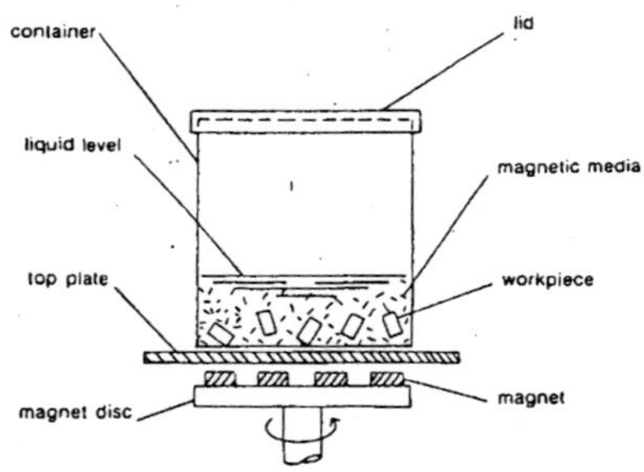


Magnetic type of non-contact centrifugal finishing

Magnetic media in various shapes can be used along with burnishing solutions to planish the work pieces to produce a highly reflective burnished finish. Magnetic shapes can also be mixed with various forms of pre-charged dry media, magnetic force is in turn used to drive one type of media against another. Pre-finishing, and in stages, finishing the work pieces, employing a combination of techniques. Methods which could possibly used in processing prostidontics, prosthetics, and ceramics.

# METALS--GOLD & SILVER

## ROTARY

| MEDIAS | PRE-FINISHING CASTINGS<br>Gold or Silver Investment Cast |
|---|---|
| Pre-charged Cones, Triangles Available in Coarse (Black), Medium (Green) & Fine (White) | 30% to 40% Media by Volume, balance...work pieces. |
| LIQUID | To Cover Load...Use Distilled water |
| ADDITIVES | Liquids available to remove Fire Scale, Brighten finishes, and Promote tumbling action. Use as directed by manufacturers. |
| RUN TIMES | 1 to 8 hours depending on surface finish of the work pieces. Check progress hourly. Finish should exhibit uniform, smooth, even texture, free from blemishes. Remove when no further improvement is evident. Do not allow liquid to turn dark, change frequently to avoid work discoloration. |

Larger work pieces may require higher ratios of media to load, or if the shapes are intricate use special types of medias to process the configurations. Larger flat areas may require longer running times, compare to rounded shapes. Check projections closely for excessive stock removal and above all, do not allow impingement of the work pieces, one against the other.

# METALS--GOLD & SILVER

## ROTARY-VIBRATORY

| BURNISHING SHAPES | FINAL BURNISHING CASTINGS<br>Gold or Silver Investment Cast |
|---|---|
| Carbon Steel and Stainless Steel: Balls, Ball Cones, Pins, and mixes | 30%--50% burnishing medial by Volume...balance of load, work pieces. |
| WATER—— (ROTARY) | To cover load...(Distilled or bottled) |
| WATER–(VIBRATORY) | 1--3 Ounces per lb. of load. |
| BURNISHING SOAP | 1/4 TO 1/2 Ounce per lb of load, use sparingly (follow manufacturer's directions) |
| ADDITIVES | Liquid types are available to maintain bright surfaces. and promote tumbling action. Use per manufacturers inst. |
| RUN TIMES ROTARY | 1-12 hours (Check frequently) |
| RUN TIMES VIBRATORY | 1-8 hours ( Check hourly) Check load often for impingement of work pieces. Adjust to larger ratio of media to load should this condition become evident. |

Intricate shapes or larger work pieces may require greater proportions of small pin shaped media to finish details. Jewelry mixes of various shapes produce good results. Stainless steel media requires far less maintenance.

# THE AUTHOR

## EDWARD E. SMITH
## 1929 -

Edward E. Smith became interested in rocks & gemstones in 1933 at the age of 4. Accompanying his parents on vacations to Grand Marias on the north shore of Lake Superior, the discovery of Agates while playing on the beaches and wading in the cold waters of Lake Superior sparked a fascination which was to become a lifetime pursuit.

Mr. Smiths father, the late E.N. Smith [1896-1981] founded Geode Industries in 1936. Served as development chairman for Geode State Park in S.E. Iowa, 1935-1951), Helping in the business as a youngster during the 1930's, the author's interest continued as mail order for Geode Industries was established. The firm became the only specialized Geode business, filling orders for Geodes, Geode Ash Trays, Bookends and Paperweights world wide.

Entering the military in 1946 at the age of 17, and during two years in the Army Special Services Div., his interest in Lapidary developed. In 1948, while a student at the State University of Iowa, this interest was cultivated when Mr. Smith was befriended by Eugene W. Gray of Iowa City Mr. Gray, owner and operator of Gray's Hardware Store (who was attested to be Iowa's oldest lapidary), passed on knowledge of a lifetime to his young protege. Free evening hours were enjoyably spent absorbing the knowledge and experience obtained over several decades by this self taught and extremely well versed lapidarist.

During the next two years the two developed a prototype of the semi-automatic diamond slicing saw for use in the lapidary.

# THE AUTHOR

Because Mr. Gray's obsession was Australian Opal, the new diamond charged saw was quite effective, compared to the "mud" saw. He also developed a single in-line lapidary bench with all sawing, grinding, sanding and polishing wheels operating at correct speeds, but with a single motor. The author's inventive course was thus established and would continue for over 45 years.

He graduated in 1952 from the State University of Iowa, with a B.S.C. degree, (Business-Science-Commerce) and returning to New London, Iowa to establish a home with his wife Lois. Interest in Lapidary grew, along with their family of four children, all of whom were destined to become involved in the family business.

He built his first tumbling machine in 1952 from home made parts and an airplane tire, and this pursuit would last over four decades of inventing and development of lapidary and industrial tumbling machines. Trade shows began in 1954, and over 500 displays were made over the next 40 years, including three trade missions abroad in 6 foreign countries, promoting the company's tumbling equipment and products.

In 1958 he developed and patented, the first seamless, off-set, polyhedral tumbling barrel, and the Viking® Tumbler introduced to the lapidary industry reduced tumbling times by 50%, from the normal 4-5 weeks, downward to 15 to 20 days. In 1962 the Vibra-Sonic®----the first vibrating tumbler in the lapidary trade---was developed and patented. Vibra-Dry® Compounds were also developed at this time. The author pioneered and introduced vibratory finishing into the Gem Trade & Industry.

Two additional rotary barrel tumbling machines were developed, a multi-compartment 115 lb. capacity, and a 300 lb. capacity single barrel unit, used in experimental processing of over 2 tons of materials during the next 3 years. Reducing tumbling times as the goal, it was soon realized that a new method must be developed. Vibration seemed the direction to pursue.

1962 saw the development of the Vibra-Sonic® tumbler, using vibration to provide rotation, reducing tumbling times to 5 to 7 days, with interchangeable containers on a universal base, with capacities from 4 to 55 pounds. The Vibra-Sonic® is still one of the most popular vibratory finishers in use, over 30 years later. As inflation grew in the 60's and early 70's a more economical unit was needed.

Early in 1974 Mini-Sonic® Gem & Jewelry Processors, were developed. A combination of rotary and vibratory action, electronically powered--with only two moving parts, speed of rotation and vibration electronically dial controlled and employing a new tumbling process so unique that its methods and process were patented in the U.S. and 5 foreign countries, receiving the first Japanese patent of its kind. Interchangeable single and multiple hoppers, with capacities from 4 to 15 pounds, hand cast solid vinyl hoppers, A.B.S. housings. A new "family" of processors!

Over the next 20 years Viking®, Vibra-Sonic®, Mini-Sonic® tumblers and Vibra-Dry® would be destined to provide new methods of tumbling not only to the Lapidary Industry, but to the Industrial, Commercial and Medical fields. Processing a wide variety of gem materials, pre-forms, cabochons, metals, ceramics, organics, industrial parts and medical implants, including heart valves.

Photo by Steve Smith

Seminar graduates pose for group photo.

Photo by Steve Smith

Lapidary plant manager Bill McCabe demonstrates polishing techniques using Vibra-Sonic® Tumblers to seminar participants, from France, Australia, and the U.S.A.

# TUMBLE ROCKS INTO GEMS

# ROTARY AND VIBRATORY TERMS

**Abrasive:** A type of hard grain used to reduce surfaces. Usually a granular form and mixed with water & materials for use.

**Abrasive Compound:** Abrasive grains usually mixed with other additives or carriers to form pastes, liquids, and slurries.

**Alkali Compound:** A mixture (usually in liquid or slurry form) designed to chemically keep metal parts clean during processing.

**Barrel:** The container (usually round or hexagon shape) which hold the materials and rotates for processing.

**Brightening Solutions:** Liquids additives (usually on the acid side) used to retard darkening of metals,

**Burnishing:** A process by which surface appearances are improved by additional processing. Also a final polishing process for metals.

**Burnishing Media:** Usually hardened polished carbon or stainless steel balls, balcones, diagonals, or pins, used in final polishing of metals.

**Burnishing Compound:** The additives used along with media to promote tumbling action and polishing.

**Compound:** Any type of additive used in the finishing of Gem Materials, Metals or Plastics..

**Cut Down:** The operation prior to polishing to radius, or pre-pare metals for the final polishing cycles.

Carrier: Any type of granule or shape used to carry abrasives.

# ROTARY AND VIBRATORY TERMS

**Cycle:** A period of processing time usually referred to as a step.

Deburring: Also termed as cut down or removing projections on metals.

**De-Scaling:** An operation used to remove fire scale from metal parts or castings, prior to pre-finish and polish.

**Diagonals:** Small rounded metal cylinders with angled ends. Also available in hardwood used in final polish.

**Fines:** Small particles of metal separated from larger materials which can oxidize during processing and darken solutions and work pieces.

**Fracturing:** Splitting or surface damage usually occurring from unbalanced loads. Usually as slurries get thinner.

**Grading:** A process of sorting materials for size and quality.

**Grinding:** Removing stock from materials, smoothing and blending surfaces in preparation for finishing.

**Impinging:** Contact of one surface against another, with enough force to cause a change, usually denting.

**Lodging:** Usually refers to media becoming stuck or wedging in detail of metals, or gem materials.

**Luster:** The reflective appearance of the objects surface.

**Mass:** The total volume of the entire contents of the load.

**Media.** Any type of material used to separate of carry abrasive.

## ROTARY AND VIBRATORY TERMS

**Organics:** Natural materials formed from animal or vegetable matter.

**pH:** Denotes the degree of acidity or alkalinity of the solution.

Peening: Using steel shot or ceramic media to fold, blend or to reduce the high parts of metals.

Rotary: A method of tumbling also referred to as barrel finishing. Also a type of tumbler employing a rotating barrel.

**Stones:** Gemstones or gem materials formed from minerals.

**Tumbling:** A generic term used to describe both rotary and vibratory finishing processes. The mass finishing of many materials in one container through various cycles or steps.

**Tumbling Barrel:** A term used to denote any type of container used in rotary tumbling or finishing.

**Vibratory:** A method of mass finishing using vibration as the driving force( as opposed to rotary) Also referring to a type of tumbler (as opposed to barrel)

**Vibratory Finishing:** A method of mass processing loads of materials. Usually the container is almost stationary, either mechanical or magnetic forces are employedto continuously rotate the contents or load.

**Wedging:** The lodging of small pieces of media into detail of metals or gem materials.